ERAN GILAT

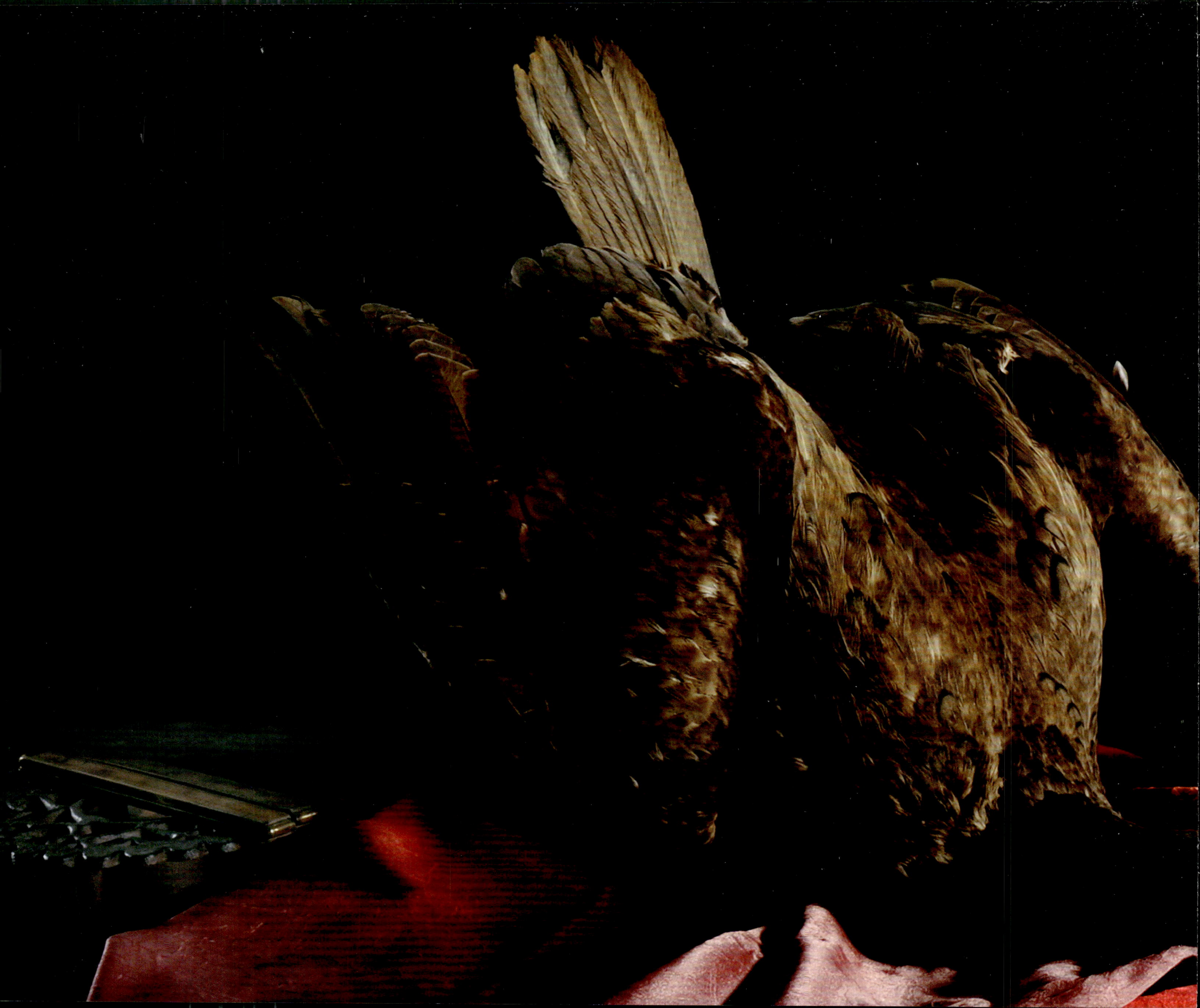

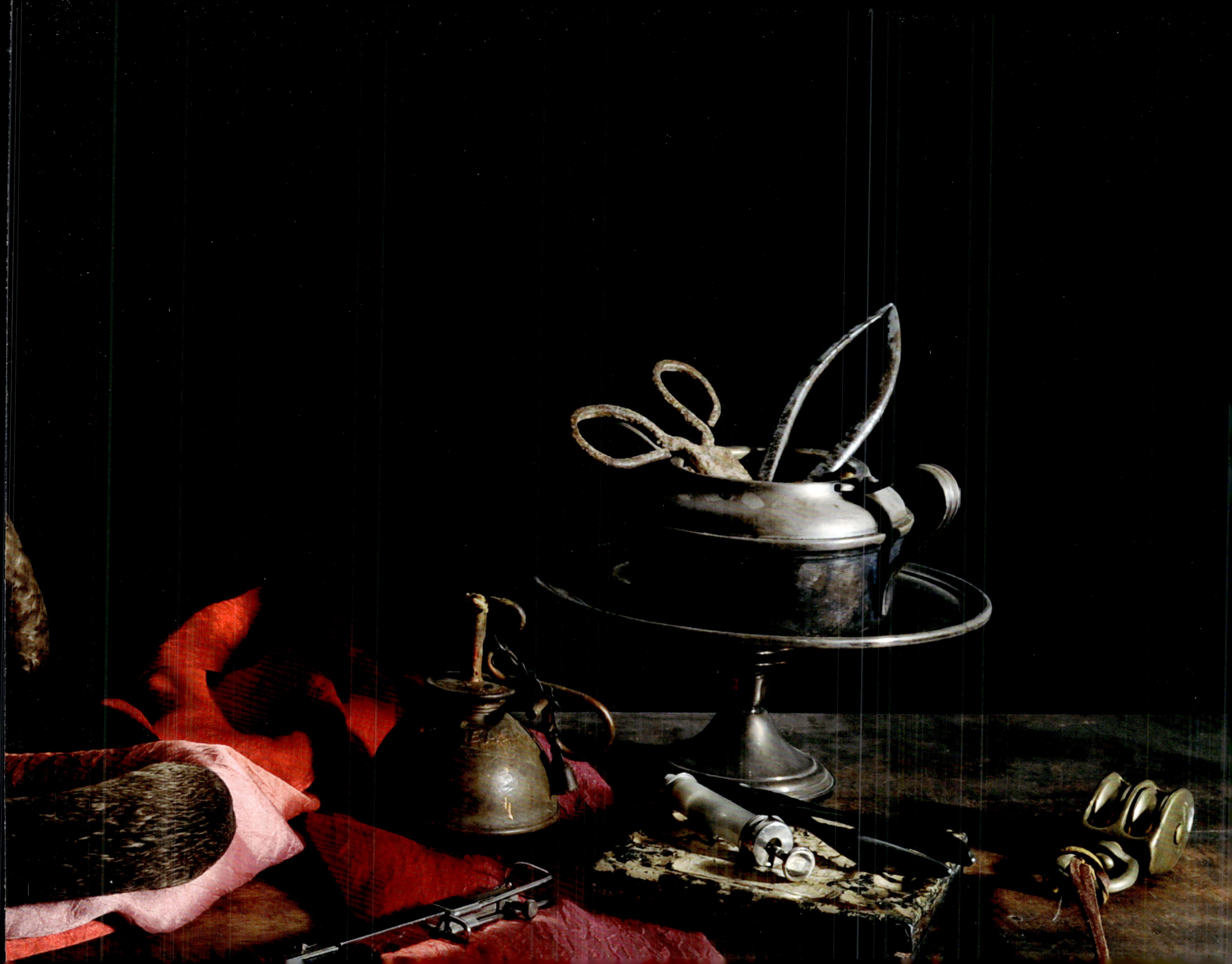

ERAN GILAT

Life Science

KEHRER

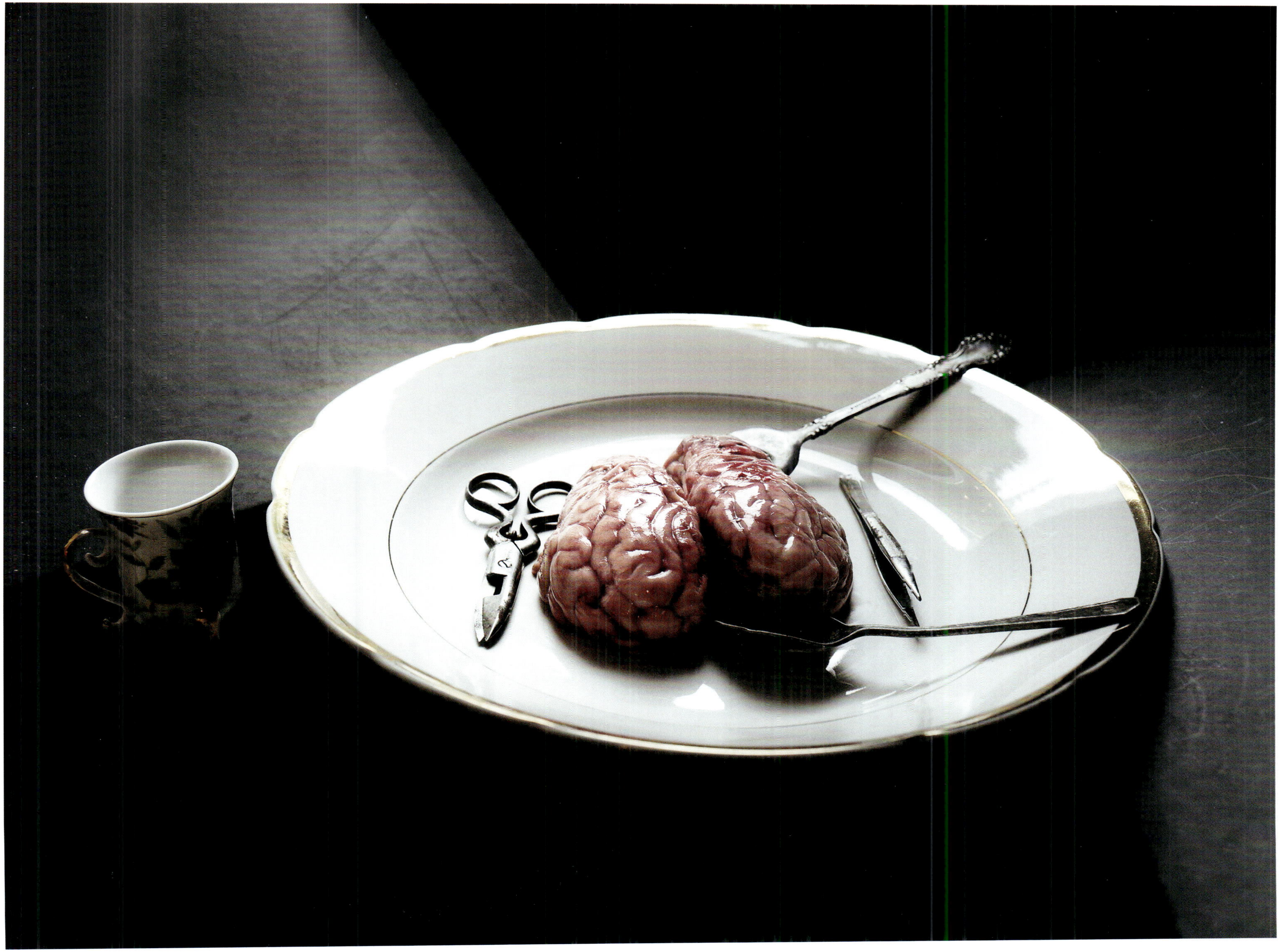

THE BRAIN AND THE VISUAL ARTS

An Emphasis on Animal Nature Visuals

Eran Gilat

As a neuroscientist engaged for many years in brain research and an avid fine art photographer, I will make an effort in this review to shed light on the interface bridging brain neurophysiology and the visual arts. I will present certain observations and opinions from the field of neurophysiology, which may directly relate to the way we experience visual art. This essay is not a scientific composition in all regards; it negotiates certain scientific findings with assumptions, hypotheses, and reflections. It aims to focus on certain aspects of neuroaesthetics, a novel sub-discipline of empirical aesthetics. This essay will negotiate the issue noted above with emphasized cautiousness with regard to our miniscule knowledge of the brain's high functions. Certain thoughts expressed here are my own, while others are from eminent scholars working in this field. A few essential scientific findings in brain research by distinguished scientists related to the topic of this essay will also be presented here.

My artwork presented in this photographic monograph is focused primarily on still life images of bodies, organs, flesh, and fauna physiognomies, all related to the human and animal arenas; models are occasionally engaged, whereby no human tissue is used and all specimens were derived from meat markets or museum collections. I believe that these genres may serve as appropriate candidates for a discussion of neurophysiology and aesthetics. My private work will also be contemplated by works taken from the history of art, as well as by a private point of view in the form of a brief and separate artist statement. The issue of "animal nature," which is emphasized in certain works of mine, will be discussed, but the reader will realize that most of the findings are relevant to the common way we analyze works of visual art.

Unfortunately, although essential revelations in brain research have been achieved, we still lack critical knowledge relevant to emotions, cognitive processes etc. These impose modesty and cautiousness when relating observations to the way we scrutinize visuals, avoiding definite assumptions and conclusions. As already stated, the field of neuroaesthetics that encompasses the issues discussed here is a novel one, engaging the interface bridging neurophysiology and the art world.

Since many images in my photographic work depict various inanimate tissues and taxidermies, confusion between animate and inanimate can be triggered, leading to "animal nature" sensation. Allow me before I continue to outline here the meaning and possible consequences of a confrontation with "animal nature" cues in the visual arts. In their review, "Disgust: The body and soul emotion in the 21st century," the psychologist Paul Rozin and his colleagues list several categories of classical elicitors of disgust: Among these is *Core Disgust*, which basically deals with the oral incorporation of food, which is highly subject to cultural practices.[1]

The authors present a second category, which is the most relevant to this essay: *Animal-nature Disgust*, encompassing four kinds of elicitors, namely poor hygiene, body envelope violations, contact with death as touching or visualizing a corpse, and inappropriate sex (as with animals or siblings). I will extend by including visualizing flesh, various tissues, and taxidermies, i. e. everything that reminds us of our animal origins and the fact that we are not immortal and terribly exposed to poor hygiene and body envelope violations. These raise the human major threat, endangering our vast motivation and aspirations for the establishment of an immortal route. The two other disgust elicitor categories noted in their study are "Interpersonal Disgust and Socio-moral Disgust." Socio-moral issues are frequently the dominant ingredient of many artworks, and visual cues, mostly by the press regarding regions engaged in conflicts. These may act as intriguing and captivating stimuli for the viewers, since they combine the animal nature cue and conceptual moral contemplations.

One of the pioneers contemplating the interrelationships between art and science is the distinguished scientist Semir Zeki. Zeki's contribution to the basic understanding of the central visual system is instrumental, since he is one of the leading contributors to our understanding of the way the visual system handles motion and colors. In his book *Inner Vision: An Exploration of Art and the Brain*, Zeki attempts to unravel certain aspects of the interphase bridging our knowledge of neurophysiology and the arts.[2] His composition is nourished by scientific revelations and philosophical thinking and contemplations. Zeki is responsible for a significant but highly debatable statement regarding neuroaesthetics: "[...] artists are neurologists, studying the brain with techniques that are unique to them and reaching interesting but unspecified conclusions about the organization of the brain [...]." He goes on to quote Henri Matisse: "[...] the artist will seize [a truer, more essential character] so that he may give to reality a more lasting interpretation." Zeki supports his opinion by the fact that "[...] they create images that please their brains, and hopefully will cause other brains to enjoy its aesthetics as well." "The function of art," he says, "is therefore an extension of the function of the brain." A similar statement by Zeki is also discussed briefly by the highly esteemed philosopher John Onians in his book on neuroarthistory, in which he paraphrases Semir Zeki: "[...] the artist is in a sense, a neuroscientist, exploring the potentials and capacities of the brain, though with different tools. How such creations can arouse aesthetic experiences can only be fully understood in neural terms. Such an understanding is now well within our reach."[3]

Onians continues: "I say outlines because our knowledge of how the brain works is still only very sketchy and we are therefore only able to account for how we see in an imperfect way. Given this imperfect knowledge, it is even more difficult to say much, if anything at all, about how and where the aesthetic experience that a work produces arises, nor yet about the neurology underlying the emotional experience that it arouses."

Like Zeki, I see art as an "extension of the artist's brain," but this holds true for most of our activities, since the brain dictates all our endeavors and, in the case of the artist, creation of art is imposed by his brain. Personally,

1
Paul Rozin, Jonathan Haidt and Clark McCauley, "Disgust: The body and soul emotion in the 21st century", in: Bunmi O. Olatunji and D. McKay (eds.), *Disgust and its disorders. Theory, assessment, and treatment implications* (Washington, D. C.: American Psychological Association 2009) pp. 9–29.

2
Semir Zeki, *Inner Vision: An Exploration of Art and the Brain* (Oxford: Oxford University Press 1999).

3
John Onians, *Neuroarthistory: From Aristotle and Pliny to Baxandall and Zeki* (New Haven, Connecticut: Yale University Press 2008).

I feel that the artist may bear certain skills of a neurophysiologist, but I believe has more of an expressive patient than of a genius neurologist, planning how to arouse his spectator. In many cases, artists express their inner visions, states of mind, subconscious archives of joy and trauma, protest, raising moral issues, and playing around with their aesthetic ideas. The quality of the art will depend on the "story" the artist is telling us through his work. Some will present boring materials and others will titillate, intrigue, and arouse us with their presentation. I tend to think that their inner world is entangled with disquieting and intriguing elicitors, and we are gifted to face it. In reference to these, I wish to briefly quote and discuss the link suggested regarding the work of the Surrealists and "automatism expression," as well as the possible revelation of the unconscious, since it may be especially relevant to photographers focusing on semi-spontaneous activity, which is partially expressed in this photographic monograph.

Adina Kamien-Kazhdan states in the exhibition catalog to *Surrealism and Beyond* in the Israel Museum (2007): "[...] artists associated with the surrealist group developed automatic techniques, which served as its visual counterpart. These techniques were used to intensify the artist's visionary faculties and allow expression of one's inner self. This culminated in multiple automatic techniques, including Andre Mason's *'automatic drawings,'* Joan Miró's semi-automatic works, and Max Ernst's *frottages* and *grattages* of the 1920s. The unconscious forces characterizing the automatic techniques occur typically in two stages. At first the artist suppresses conscious control, and by means of a rapid process exploits the chance to create a free ambiguous form or texture. In some cases the artist leaves the stimulus as is, without further artistic intervention. In some cases aesthetic judgment leads to further late alterations."[4]

Since most of these issues concern high functions of the brain, we are still at a remote position to make definite assumptions and statements. Numerous research projects of our visual apparatus from the retina to the thalamic Lateral Geniculate Body (LGB), to the striate cortex visual regions and further to association cortices, resulted in important neurophysiological findings; unfortunately, however, insufficient information is available at the moment for the understanding of what makes an artwork a "visual miracle" for a certain viewer, and how to clearly differentiate between art and other primal arousal cues. The field of neuroaesthetics is initiating preliminary research into this issue, using the accumulated, novel findings together with logical analyses.

What can we derive for our purpose from recent scientific studies of the brain?

I consider certain studies of the brain, such as studies on Brain Plasticity and certain functional Magnetic Resonance Imaging (fMRI) studies designed to explore brain activity while viewing visual cues such as artworks, faces etc. under the umbrella of the field of neuroaesthetics. One category of these studies engages plasticity (shaping) of our brain's visual neural machinery.

Visual Plasticity studies:

Most brain plasticity studies naturally focus on the young newborn animal, i.e. on early life stages of visual exposure (mostly post-natal). I will, however, also discuss encounters with intriguing visuals that may affect our brain reaction to visuals and lead to accommodation. Plasticity research engages the effect of visual stimuli on the

4
Adina Kamien-Kazhdan, *Surrealism and Beyond in the Israel Museum*, exh. cat. Israel Museum, Jerusalem (Jerusalem 2007).

build-up and shaping of the brain's neural circuits. The focus in many of these studies is naturally on one sensory modality, namely the visual apparatus, although other regions are also affected. I consider this section essential to our discussion, since I believe that our attitude to what we see highly depends on our former visual experience, in early and later stages of our lives, in addition to our genetically predetermined "visual brain." This may directly relate to art evaluation. Cultural and environmental attributes enforcing their effects on the cognitive connotations involved are also essential. The understanding of the mechanisms of post-natal plasticity is vital for extending our knowledge regarding the shaping of the visual system in humans, from the newborn to the adult viewer. Naturally, we enrich our newborn surroundings with various colorful cues, sometimes in motion. Should we? Out of numerous studies by distinguished scientists, such as David Hubel and Torsten Wiesel (both Nobel Laureates), Collin Blackmore, and others, I choose to sample a state of the art study by Tanaka et al. (2009)[5], focusing on plasticity during the post-natal critical period for cue orientation plasticity in the cat visual cortex.

Orientation selectivity of primary visual cortical neurons is an important requisite for the shaping of visual perception (selective neural response of a neuron to a stripe cue in a certain orientation with lack of response in other orientations). Although numerous studies have been previously devoted to the question of how orientation selectivity is established and elaborated in early life, how the susceptibility of orientation plasticity to visual experience changes in time remains unclear. In this study, the authors showed a post-natal sensitive period profile for the modifiability of orientation selectivity in the visual cortex neurons of kittens reared with either horizontal or vertical stripes orientation exposure. Two weeks of over-representation of the exposed orientation, and two more weeks of consolidation, altered the normal visual orientation maps, establishing a larger population of neurons with selective preference to the post-natal exposed orientation. Selective orientation exposure that started later than fifty days post-natal in kittens did not change normal visual neurons orientation maps. The study revealed plasticity and consolidation of orientation selectivity in a critical period of early life. In summary, visual exposure in early life stages, while shaping the plastic visual system, may affect the mature central vision machinery and hence the establishment of an individual preference for certain cues, possibly later on for certain genres of visual cues.

Late life exposures and accommodations to intriguing and disquieting visuals are also well known. We know that in communities of hunters, the young will not hesitate to remove the skin of a dead animal. Medical students face rough times while studying anatomy. Clinical staff and scientists learn to deal with "animal and human nature cues," including body envelope violations, and the exposure is less traumatizing over time. Cultural atmosphere and habit also play significant roles.

What can we derive for our purpose from recent fMRI scientific studies of the brain?

With regards to human fMRI studies, out of the numerous studies on neuroaesthetics, I will select one pioneering study by Hideaki Kawabata and Semir Zeki (2004), which focuses on the neural correlates of beauty.[6] The authors used the technique of fMRI to address the question of whether there are brain areas that are especially active, linked by the viewer to the consideration and

5
Shigeru Tanaka, Toshiki Tani, Jerome Ribot, Kazunori O'Hashib, and Kazuyuki Imamura, "A Postnatal Critical Period for Orientation Plasticity in the Cat Visual Cortex" In: PLoS ONE 4: e5380. doi:10.1371/journal.pone.5380 (2009).

6
Hideaki Kawabata and Semir Zeki, "Neural Correlates of Beauty", in: *J Neurophysiol*, no. 91 (2004), pp. 1699–1705.

classification of visual beauty. In their study, subjects view four categories of paintings: portrait, landscape, still life, and an abstract composition. They were asked to report their aesthetic judgment, exhibiting judgment-specific brain activity in tasks of comparisons of beautiful vs. ugly and beautiful vs. neutral (as to their own judgment, regardless of category. When searching for brain areas that responded specifically to one category of painting (e.g., portrait) over the others (portrait vs. non-portrait, landscape vs. non-landscape, still life vs. non-still life, and abstract vs. non-abstract) they found that, in the abstract case, no site specific activity was produced; still life produced the greatest change in the V3 region, portraits in the middle fusiform gyrus (fusiform face area), and landscapes in the para-hippocampal place area. Numerous recent fMRI studies were aimed at unraveling certain central aesthetic assessment issues; localization of site specific cognitive activity and certain region's cross talk is the sole finding in most of them. The regions responsible for face recognition and their expression's effects were nicely depicted by a few leading neurophysiologists (e.g. Patrik Vuilleumier, 2007)[7]. Localization of a brain-specific function merely informs us where this takes place. It does not convey deeper knowledge concerning the mechanisms underlying the cognitive process.

Localization of vital areas, such as primary motor or sensory and visual cortices, the Broca's and Wernicke's speech areas, could needless to say be critical knowledge when a neurosurgeon accesses the brain for the removal of a tumor or an epileptic foci; but what do we really study about higher cognitive functions, such as the "art and science" interface, in the cognitive aspects? Unfortunately, very little at the moment, if at all.

I am thus left to contemplate interesting and rationally intuitive theories by Semir Zeki, Vilayanur Subramanian Ramachandran, John Onians, and other scholars, who are experts either in neurophysiology, neuroarthistory, or philosophy in general (mind you, scientists are devoted to experimental data, while philosophers may contemplate and theorize). At the moment, we have to rely primarily on theories and speculations while we patiently wait for breakthroughs in the research of higher brain cognitive functions. Some claim that it is beyond our reach. Let us thus try to negotiate the reason why. In the words of Anjan Chatterjee: "A person stands transfixed before a Mark Rothko abstract painting [it could also be a captivating photograph or any other work of art, E. G.] oblivious to everything else. Embedded in this scene are questions that strike at the very heart of this talk. What is the sensation being experienced?"[8]

I will not deal here with optical phenomena affecting our perception of visual art; there are many optical maneuvers to enhance visual satisfaction by the viewer. I will focus on aesthetics, content, and message, be these neutral or disquieting and intriguing, with my innate feeling that a good work of art should be captivating and bear certain arousal stimuli in the viewer's mind. I would like to favorably adopt here a principle suggested by Ramachandran and Hirstein (1999) in their excellent essay on "The Science of Art. A Neurological Theory of Aesthetic Experience."[9] The authors suggest that "the artists either consciously or unconsciously deploy certain rules or principles (laws) to titillate and ignite the observer brain." One of their suggested visual enhancers is the "peak shift effect" principle, a terminology well documented, borrowed from animal behavior tests. The more the award-associated visual cue is *enhanced*

7
Patrik Vuilleumier, "Neural Representation of Faces in Human Visual Cortex: the Roles of Attention, Emotion, and Viewpoint", in: Naoyuki Osaka, Ingo Rentschler, and Irving Biederman (eds.), *Object Recognition, Attention, and Action* (Tokyo: Springer Japan 2007), pp. 119–138.

8
Anjan Chatterjee, "Neuroaesthetics. Researchers unravel the biology of beauty and art", in: *The Scientist*, May 1, 2014: http://www.the-scientist.com/?articles.view/articleNo/39802/title/Neuroaesthetics/, last accessed on May 9, 2016) and idem, "Visual Art", in: Jay A. Gottfried (ed.), *Neurobiology of Sensation and Reward* (Boca Raton, Florida: CRC Press 2011), pp. *391–401*.

9
V. S. Ramachandran and William Hirstein, "The Science of Art. A Neurological Theory of Aesthetic Experience", in: *Journal of Consciousness Studies*, vol. 6, nos. 6–7, 1999, pp. 15–51.

and *exaggerated*, the more it is preferred and selected by the animal. The authors mention the fact that when, in a certain task, a rat learns that a rectangular shape selection out of other geometrical shapes is accompanied by a reward (while punishment follows a square choice, for example), it develops what they call a "rectangularity" preference. Thus, that the more the rib's length proportions favor a rectangular shape, the more it will be preferred over a rectangle with reduced "rectangularity". "Accentuation" and "exaggeration" are favored.

The authors present another example: the seagull phenomenon (documented by Nikolaas Tinbergen) as an enhanced response to an artificial natural visual cue, a rod with a significant red mark, a "super beak" that mimics the natural cue for babies (i. e. the seagull mother's beak).

They give these captivating, visually accentuated cues the name "super stimuli" as a synonym for the animal experimental notion, the "peak shift effect." Like the authors, I sincerely and intuitively consider the peak shift effect principle or its mate "super stimulus" as a possible potent elicitor of emotions when visualizing and contemplating an artwork. When making an effort to explain his attitude to captivating photographs, Roland Barthes (1993), coined the term "*punctus*."[10] I believe that Roland Barthes' "punctus" may serve as a certain replacement to Ramachandran's notion of a visual "*super stimulus.*" Both point to features that "make it" a potent visual elicitor "for you." My punctus is not necessarily your super-stimulus. As I stated previously, "visual miracles" are highly dependent on the individual. Genetics, past exposure, and cultural references may determine our visual and conceptual attitude preference. We differ. However, we see that certain visual images become legendary and serve as super stimulus for many viewers who appreciate its strength.

One cannot forget the Vietnam War image of the naked, nine-year-old Kim Phuc, badly burned and running for her life. That image was taken by the photojournalist Nick Ut, who was only 21 years old at the time. It jolted people around the world. Some say it hastened the end of the Vietnam War. Horror or tranquility, a "punctus" is obligatory.

I will attempt to state here that the "animal nature" cue (the essence of this photography book) may serve as a powerful stimulus, perhaps a "super stimulus" for numerous viewers. "Animal nature" is known as a primal potent trigger for the contemplation of our temporal visit to this universe. Animals in general and body envelope violations proclaim mortality, a fear and sincere curiosity provoking stimulus. In his book *Death* (2012), Yale philosopher Shelly Kagan confronts the meaning of mortality. He states: "There is one thing I can be sure of: I am going to die. But what am I to make of that fact? Once we accept this truism, questions arise: 'Are people just material objects, nothing more? Is fear of death appropriate? How should I live in the face of death?'"[11]

The philosopher Ernest Becker discusses the questions raised above in his book *The Denial of Death* (1973): "Human beings like animals are mortal, and we know it. Our sense of vulnerability and mortality gives rise to a basic anxiety, even a terror, about our situation. So we devise all sorts of strategies to escape awareness of our mortality and vulnerability, as well as our anxious awareness of it. This psychological denial of death, Becker claims, is one of the most basic drives in individual behavior, and

10
Roland Barthes, *Camera Lucida: Reflections on Photography* (London: Vintage Classics 1993).

11
Shelly Kagan, *Death* [New Haven, Connecticut: Yale University Press (The Open Yale Courses Series) 2012].

12
Ernest Becker, *The Denial of Death* (New York: Simon & Schuster 1973).

is reflected throughout human culture."[12] In her Ph.D. thesis, *Disgust and Ugliness: A Kantian Perspective* (2011), Mojca Kuplen states: "[...] a painting even when representing a discomforting subject matter, can nevertheless be a pleasurable one, while a depiction of disgusting objects in photography provokes rejection."[13] This is because the nature of the object is more sensibly presented in a photograph and thus more easily provokes our imaginative powers, on which disgust depends. In addition she suggested that "[...] if the representation of the disgusting object does not threaten its artistic image; that is, if we are still able to retain distance to the nature of the object, the aesthetic representation can remain successful. In this case, we have a genuine situation in which the visceral reaction to the disgusting object has been suspended by its presentation. Many art works illustrate that disgust's anti-aesthetic effect and our receptivity to those works can be suspended either by stylistic control or by connecting the object with certain conceptual and moral issues."

The issue of approaching the brain with, on the one hand, an ambivalent elicitor of repulsion and, on the other hand, curiosity was raised before by philosophers, who have expressed the paradoxical enjoyment and satisfaction gained by viewers of certain repulsing visuals. In his work *Observations on the Feeling of the Beautiful and the Sublime* (1764), Kant comments that "nothing is so much set against the beautiful as disgust."[14] And in the later *Critique of Judgment* (1790), he claims that what makes the object abhorrent "is precisely its outrageous claim for desirability."[15]

The writer Sianne Ngai (2005) quotes Proust's *Swann's Way*: "I thought her so beautiful that I should have liked to be able to retrace my steps so as to shake my fist at her and shout, 'I think you are hideous, grotesque; how I loathe you!'"[16]

An individual response is obviously expected, each person bears his unique visual elicitors and his unique emotional response depending on neurophysiological genetics, past-exposure experience, and cultural orientation. I believe that, in general, an attractive artwork should comprise a conceptual statement, either tranquilizing or disquieting, together with appealing aesthetics. I will go even further and use the term "miracle" when contemplating aesthetics of abstract art and state that everyone is equipped with his own central neurophysiological apparatus, which determines their visual miracles as viewers. As a scientist accustomed to parameters, measurements, and statistics with clear cut results, I have to clarify that this kind of thinking takes place within the frameworks of theory and make-sense contemplations. Nevertheless, one cannot ignore statistics or when numerous viewers are fascinated by a certain painting or photograph; we have to generously admit that it is captivating from the point of view of fine art, or as an important document, conveying a significant conceptual message.

13
Mojca Kuplen, *Disgust and Ugliness: A Kantian Perspective* (Ph.D. thesis in Philosophy at the Central European University, Budapest, Hungary, 2011).

14
Immanuel Kant, *Critique of Judgment* (1790), trans. by Werner S. Pluhar (Indianapolis/Cambridge: Hackett 1987).

15
Ibid., *Observations on the Feeling of the Beautiful and Sublime* (1764).

16
Sianne Ngai, *Ugly Feelings* (Cambridge, Massachusetts: Harvard University Press 2005).

ABOUT THE ARTIST

Eran Gilat is a neuroscientist and avid fine art photographer. His scientific research focuses on the study of the mechanisms underlying epilepsy, as well as on the development of an innovative cure for this illness. In recent years, highly inspired by his longstanding confrontation with biological tissues and para-clinical constellations for the study of neuroscience, Gilat has found himself directing vast attention to the still life and conceptual portraiture in photography. He deals with the aesthetics of the scene, improvising various contexts; the tools and paraphernalia depicted are not merely ones typically used in the clinical operating room. "It takes a while for a young clinician or researcher to adapt himself to laboratory or hospital scenes and thus perform on an adequate level," Gilat explains. Ultimately, this is achieved through extensive training; whereby some never adjust to the visuals. Gilat feels that his photographic activity also carries him to these regions. His "Life Science" project forces biological tissue into relatively pleasant, at times artificial scenarios, where he can contemplate disquieting issues of materialism, mortality, and erotica, corresponding with the complicated and intriguing category of "animal reminder" in the visual arts. Gilat's work also negotiates various tones of violence. We tend to describe violent human behavior as animal-like, beastly, and revolting, associating it with aggressive animal behavior. According to Eran Gilat, however, the animal kingdom is guided purely by rules of survival caused by a harsh and relentless process of natural selection. No one would consider the employment of essentials measures for the preservation of the species as being "cruel;" this has been emphasized by Charles Darwin, as well as by a number of eminent contemporary scholars. Gilat believes that, with regard to moral conduct, we are in many ways inferior to the animal kingdom, while at the same time being superior in terms of our intellectual competence.

He also contemplates what he calls "Brain Research Glory." As a senior neuroscientist, he has been working for many years in the field of brain research. Here, he has made great achievements with regard to the understanding of certain mechanisms underlying cerebral functions, as well as of the frustration accompanying scientists realizing the unsettling complexity of higher brain functions. Gilat also applies certain surreal approaches to the "personal glory" and satisfaction of eminent scholars with regard to their "glorious profession," combined with the modesty that results from the recognition of the current limitations and uncertainty of decoding critical high central mental functions such as cognitive functions, perception, and memory.

THE LIVING DEAD WORLD OF ERAN GILAT'S *LIFE SCIENCE* PROJECT

‹ ›

Smadar Levy*

Eran Gilat's *Life Science* series of staged nature photographs comprises beautifully organized, exquisite, and captivating images. The series is based on studio photographs of objects accompanied by living tissue from human internal organs or animals, in some cases on a heavy wooden table, either bare or covered with a tablecloth. Gilat's photographs depict staged compositions of still life motifs. The nature in his photographs is derived from the animal kingdom – a fish, an octopus, a heart, a chunk of meat, strands of hair, or an animal horn. Still life originates in the kitchen, or perhaps in the laboratory. Despite its appearance, however, the "living" in Gilat's work is already in a mortal state, and his still life objects are no longer in use. Specimens for his staged photography are borrowed from meat markets or museum collections (e.g. the Tel Aviv University Natural History Museum). Gilat builds small but meticulous scenes on tables: At first glance, a composition may appear like preparatory documentation prior to a medical or research procedure; this is, however, not the case. At other times, the composition may appear as though it would serve a cataloging process or scientific preservation; nevertheless, this, too, is not the case. In other instances, the color red is reminiscent of blood in a post-surgery scene; but this, too, is misleading. With cynical humor, Dr. Gilat, a neurobiologist by profession, converts clinical science into clinical photography. He toys with the aesthetics of the composition and creates living, sensual, and dramatic scenes—from the imaginary to the surrealistic—scenes of the *world of the living dead*.

Gilat confronts the spectator with staged photographic still life motifs. In his photographs, nature—the world of fauna—is transformed into *nature morte*. What is more, he stages his protagonists as though they were in a theater—against a black background, with a red and bright presence under focused clandestine lighting. This is, however, staged beauty, i.e. on-stage direction and baroque theatricality, which in fact alludes to the genre of still life painting.

This genre, which was widespread in Central Europe during the 17th century, focused on the beauty of everyday objects, plants, and animals, which were presented or mounted on tables in interior settings in privates homes or in the artist's studio. The focus on beauty was expressed in paintings of staged structuralism, emphasizing materiality and lighting. Here, one is reminded primarily of images of animals in *Vanitas* paintings, a symbolic and philosophical subgenre of still life paintings, characteristic of the Dutch and Flemish painting of the time. The source of the term "vanitas" (emptiness) can be found in the Latin translation of the biblical verse, "Vanity of vanities, saith Koheleth; vanity of vanities, all is vanity" (*Vanitas vanitatum omnia vanitas*)[1], as well as in the Christian perception that required believers to remember that they too must die: "*Memento mori.*" In these paintings, the images of animal carcasses, skulls, candlelight, or rotting fruit on one hand, and the beauty and freshness of the flowers on the other, constituted a reminder of the passing of time; withering, drying up, and decay were presented as a necessary *memento mori*.

In line with this perspective, the quest for the classical and primary meaning of the *memento mori* dictate becomes clear in the photographic subjects of Gilat, who clings to and scrutinizes the moment of death on either side of that moment—those whose death is near, and those that are already dead. Beauty is profoundly contemplated in Gilat's work:

> *Death is the mother of beauty; hence from her,*
> *alone, shall come fulfillment to our dreams*
> *and our desires.*
>
> WALLACE STEVENS, "SUNDAY MORNING"[2]

Gilat ponders the moment of death, and the beauty of his play, as well as the pictorial referent of his photographic work, entails an additional examination of the presence of the concept of death in the pictorial genre and, moreover, its presence in the medium of photography.

Let us now expand upon the term "still life" and "death," which is incorporated into it, as well as on the immanency of death in the medium of photography. Etymologically speaking, the term "still life" did not come into use until the mid-17th century. Its source is the Dutch word *stilleven*, which appeared in different inventories and indicates the description of an item—either natural or manmade—an "inanimate object" or "immobile nature." In 1675, the German artist and art historian Joachim von Sandrart wrote about immobile objects in painting, but it was only in the 18th century in France that the term *nature morte* (literally "dead nature") was coined as a characteristic thematic description of painting of the time, a feature which would later serve to define the pictorial genre. Early use of this term is documented in two individual texts: in a letter written by the Duke of Nemours in 1779, in which he used the new term to describe "inanimate things" (*les choses inarimees*); and in 1780, Jean-Baptiste Deschamps also used the term *nature morte* to describe the painting of "immobile objects" (*objets immobiles*).[3]

Years later, the term "still life" was coined in English in order to describe this genre. In English, the meaning of the term corresponds with "still" in the Dutch linguistic source, and with "life," it's reverse meaning, in French. In Hebrew as well, the term "still nature" is a fusion between the meanings of the term in French and English.

Whether we relate to the term "still life" in its meaning as "immobile life," i.e. as dead and lifeless nature, or merely as literally "still" life, its meaning embodies immobility and lifelessness. In other words, the concept of death is present, even before the religious dictate.

Returning to the photographs of Gilat, it is clear that not only do these works follow the classical and primary meaning of the dictate *memento mori*, present in the pictorial genre, they are also works that conceptually investigate and correspond with the concept of death, inherent in the concept of "still nature": "still life" and "*nature morte*."

The kind of death we are dealing with, however, is not only that which is photographed, but is inherent in the medium itself, i.e. photography. Since the publication of *Camera Lucida: Reflections on Photography* by Roland Barthes[4], one can no longer speak of photography without contemplating death. Barthes related to the basic connection between photography and death and the way in which photography signifies what is absent, in both time and space; "that-has-been," reality in the past sense. In other words, every photograph is the presence of what was and no longer is. "Ultimately, what I am seeking in the photograph taken of me (the 'intention' according to which I look at it) is Death: Death is the eidos of that photograph." Furthermore, Barthes expands, the flat and inanimate film presents a misleading mixture of two concepts: the real and the living. When he states that the topic of photography is real, the photographer stealthily inculcates in us the belief that it was alive, due to a deception that prompts us to attribute to the real an absolute sublime value, the value of something immortal. However, by transferring this reality to the past ("it happened"), the photograph alludes to the fact that it may already be dead.[5]

Thus, death is inherent in photography regardless of the photographed content. Consequently, when the content matter also deals with death, the deceit the photograph presents is doubled: In Gilat's work, the real that was photographed and endorses the value of something immortal is already dead. Moreover, this double deception is redoubled when the photographic structure detaches the real image from all specific aspects of time and space and, even for a moment, relies on another genre in which the concept of death is embedded.

Make no mistake; it is not the pain of loss, nor the tragedy of death that is present in Gilat's photographs. On the contrary, it is the beauty of death, the defining link between the concepts of "beauty" and "death," in which they engage. A link that endorses desires, passions, and wishes.

Smadar Levy, Artist and Art Educator, Art Curator and Art Writer, Lecturer and Tutor at Sapir College Art School, Israel.

1. *Ecclesiastes* 1:2.
2. Wallace Stevens, "Sunday Morning," in: *The Collected Poems of Wallace Stevens*, New York 1990, pp. 66–71; here, p. 68
3. Cf. Norbert Schneider, *Still Life*, Cologne 2003.
4. Roland Barthes, *Camera Lucida: Reflections on Photography*, New York 1981.
5. Ibid., pp. 20 and 82.

* THIS TEXT WAS ORIGINALLY WRITTEN IN THE CONTEXT OF ERAN GILAT'S SOLO EXHIBITION AT THE REHOVOT MUNICIPAL ART GALLERY, ISRAEL IN 2015.

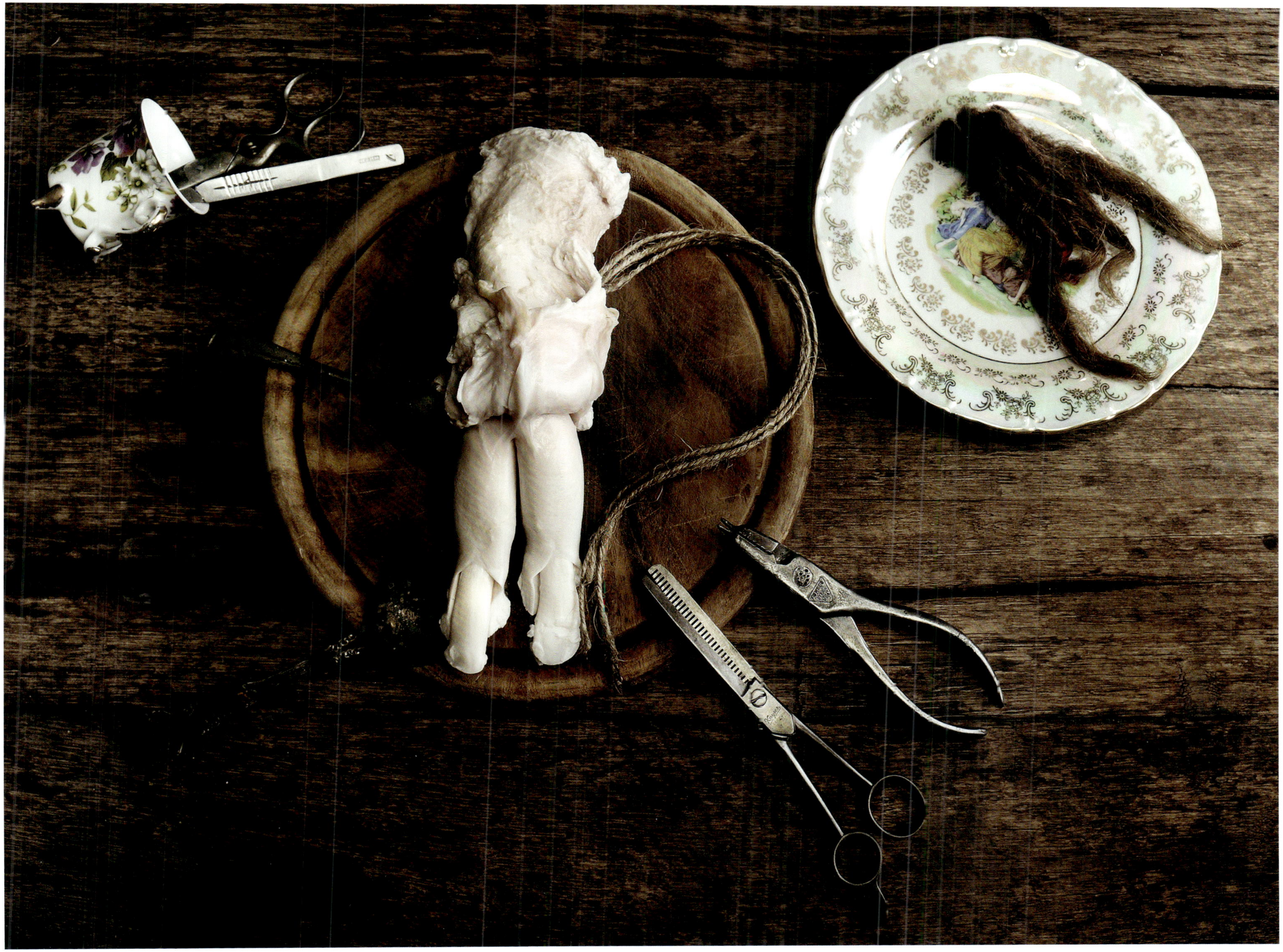

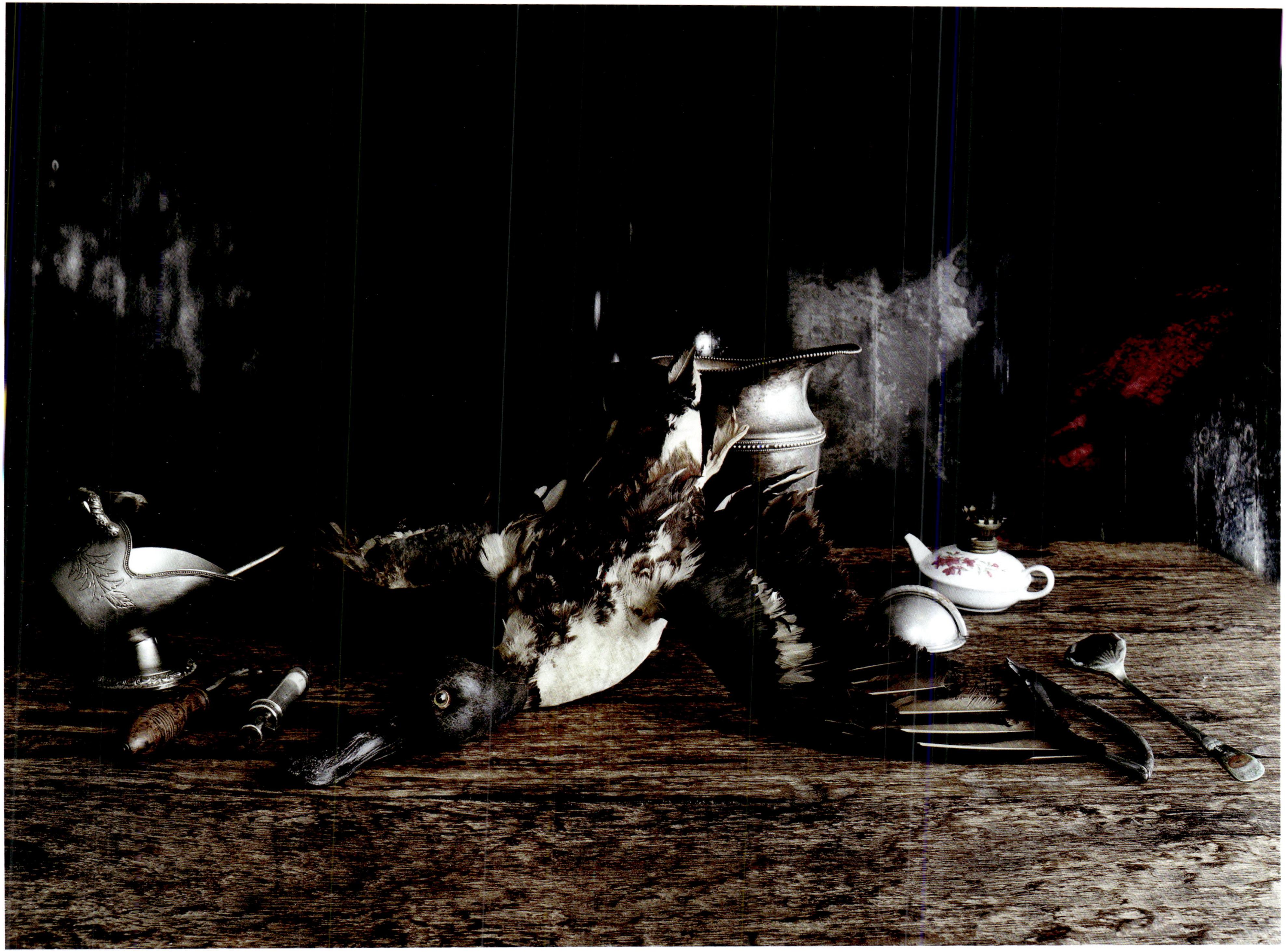

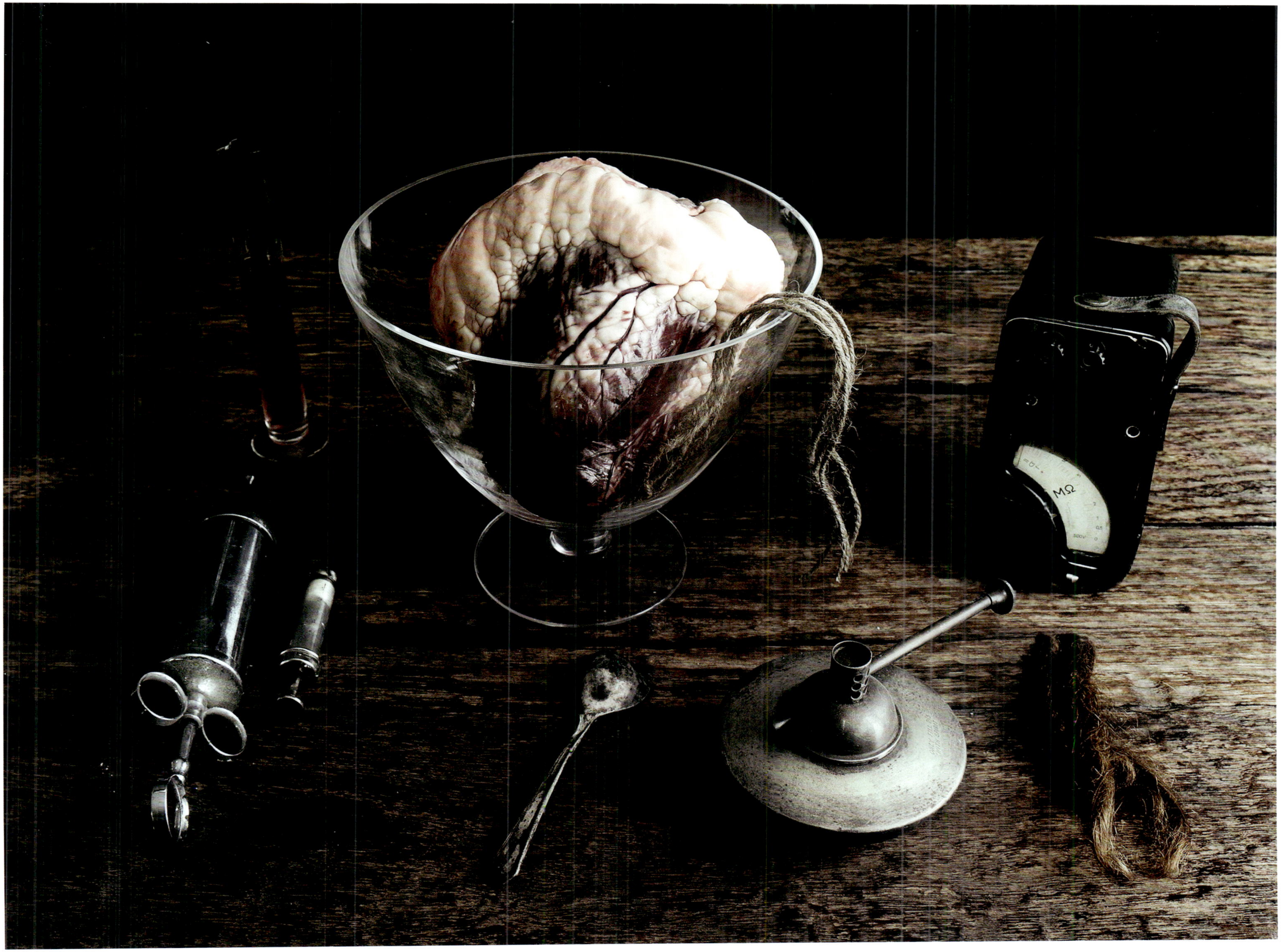
MΩ

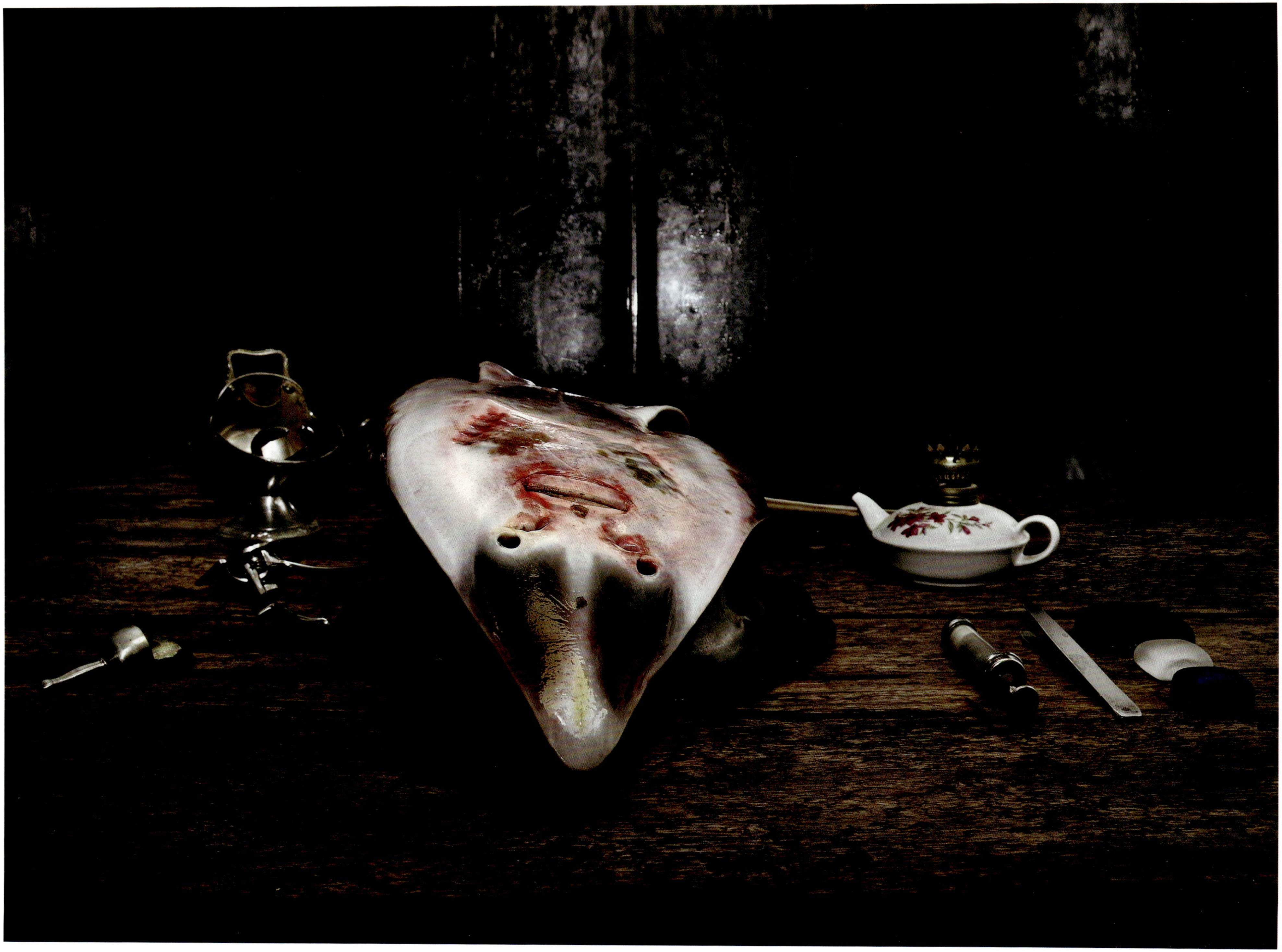

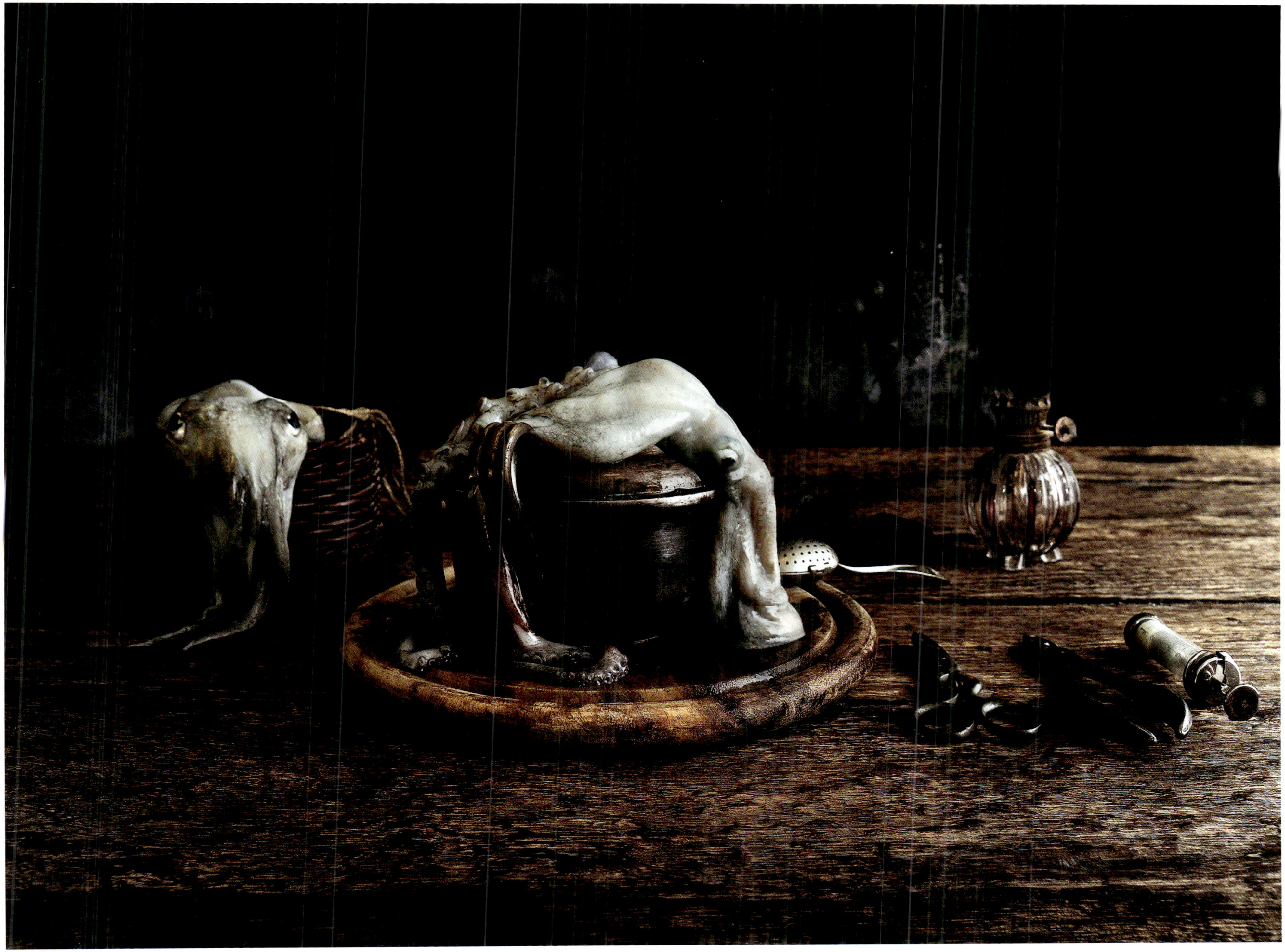

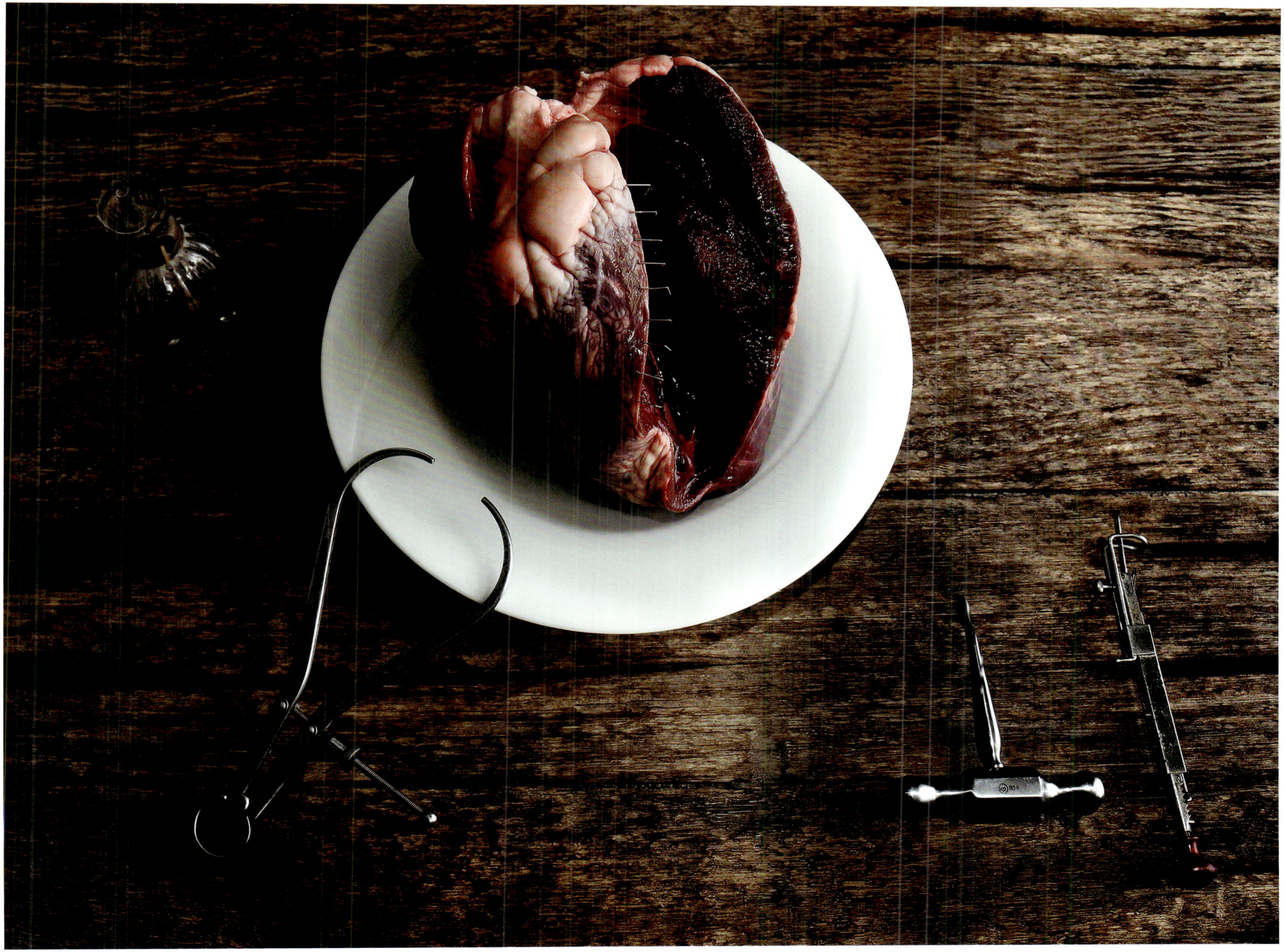

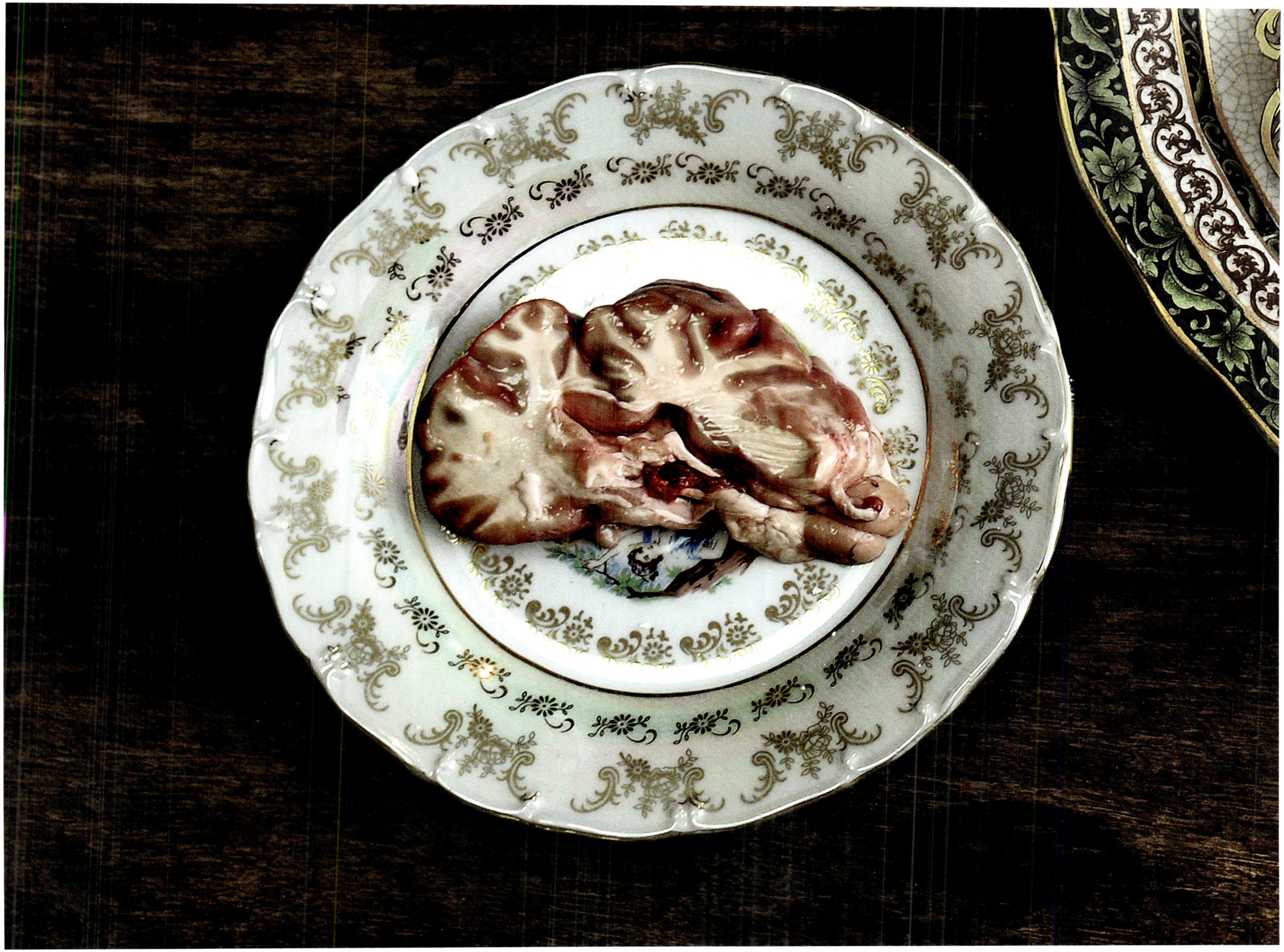

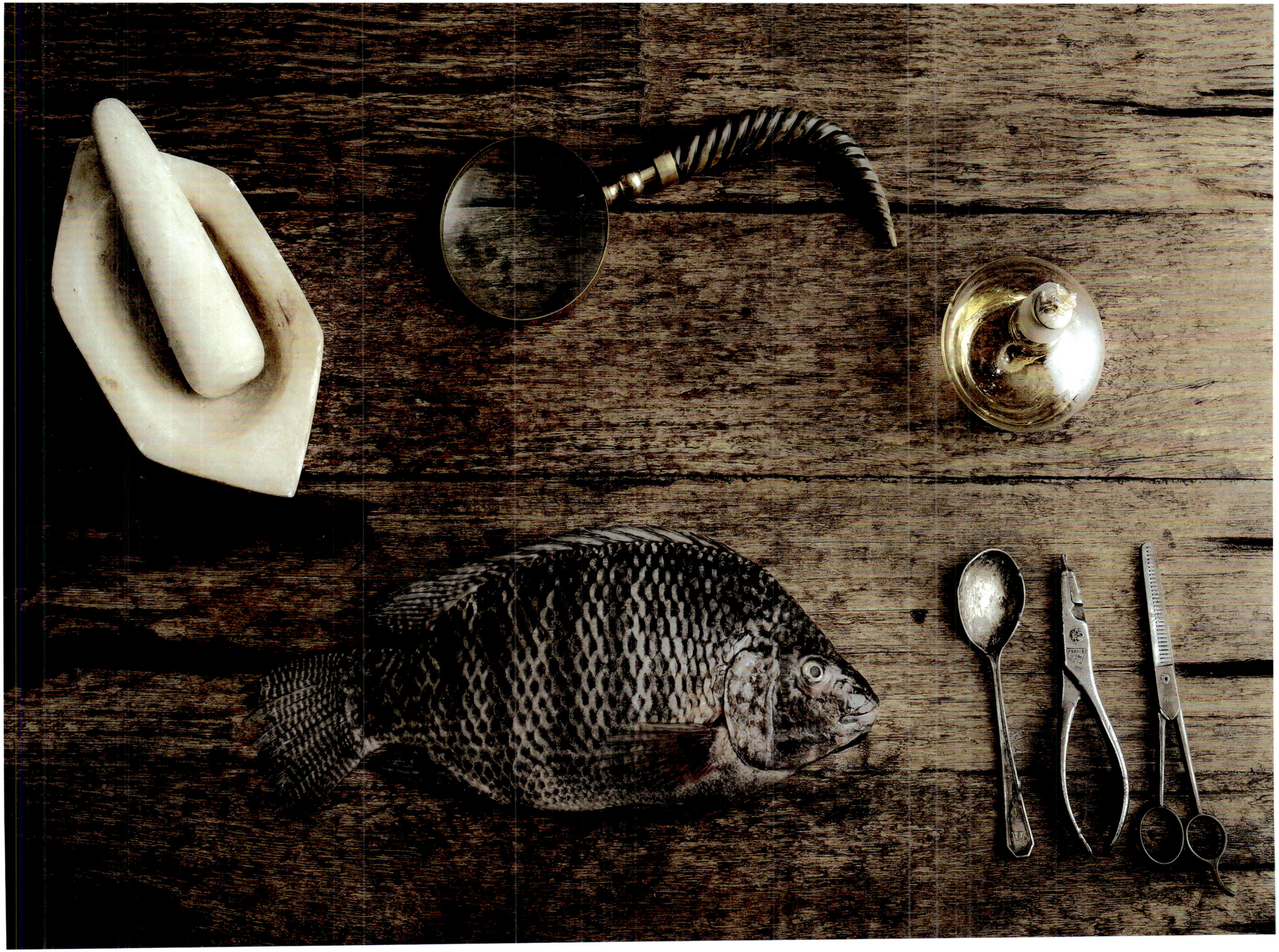

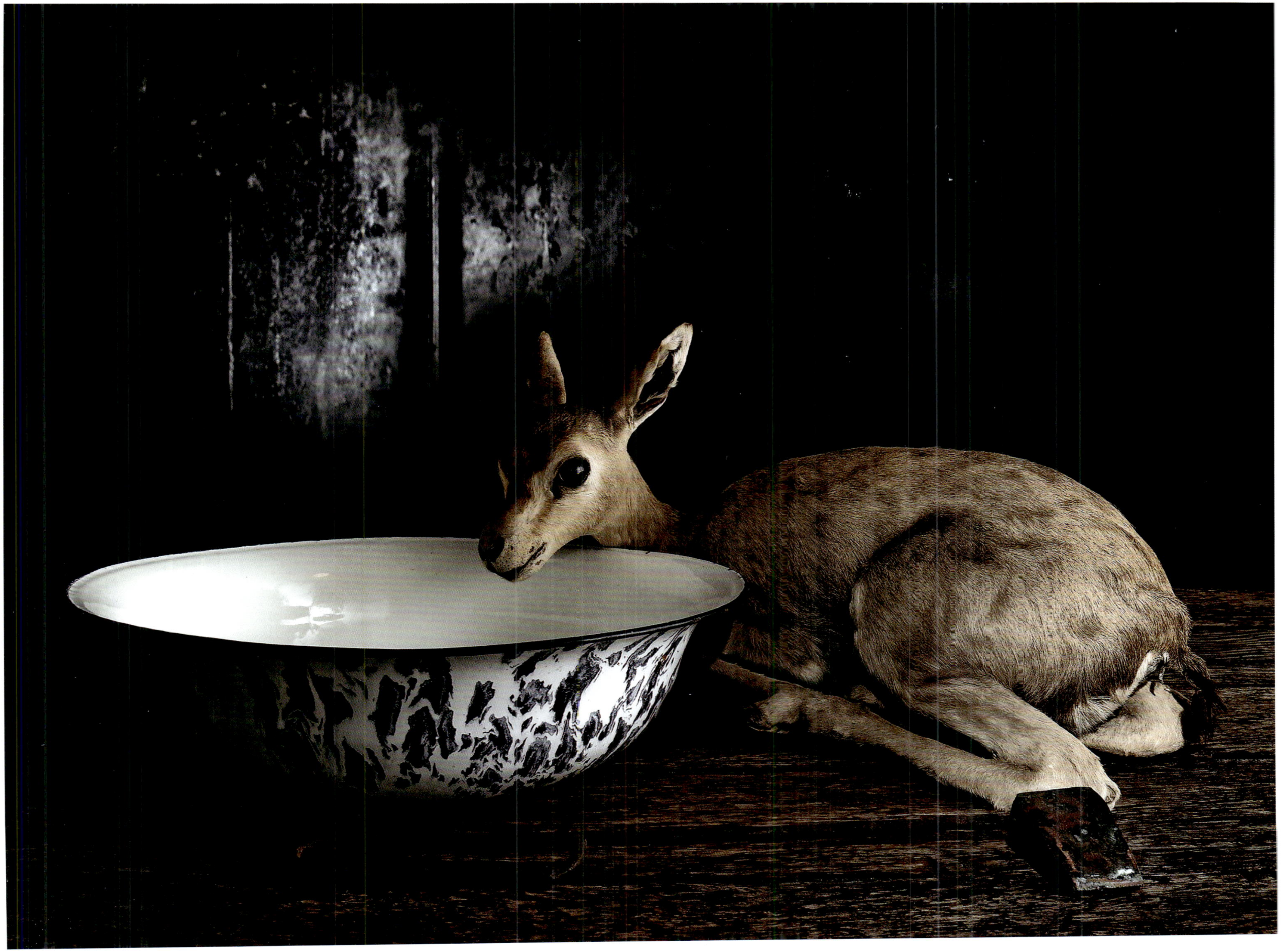

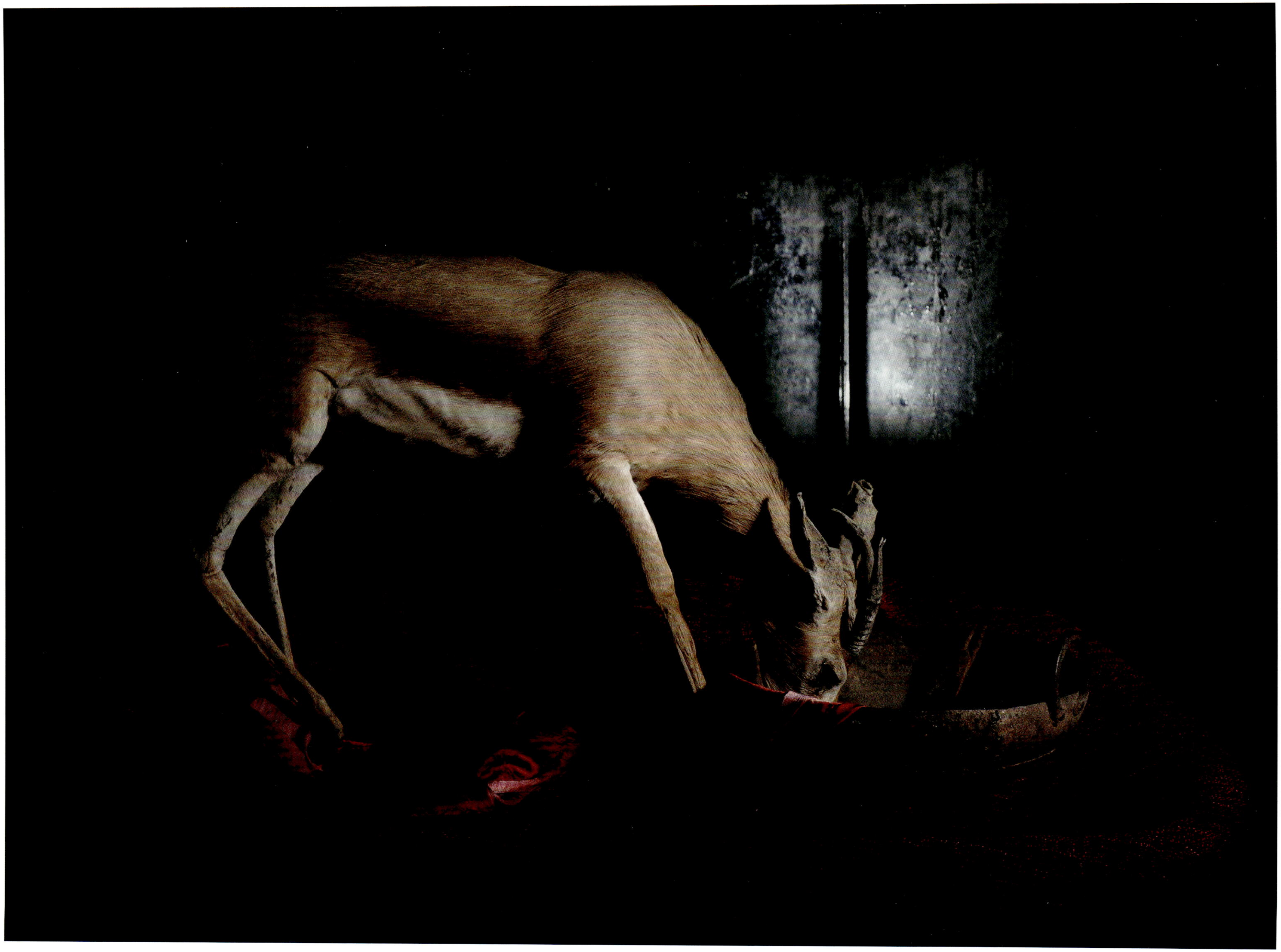

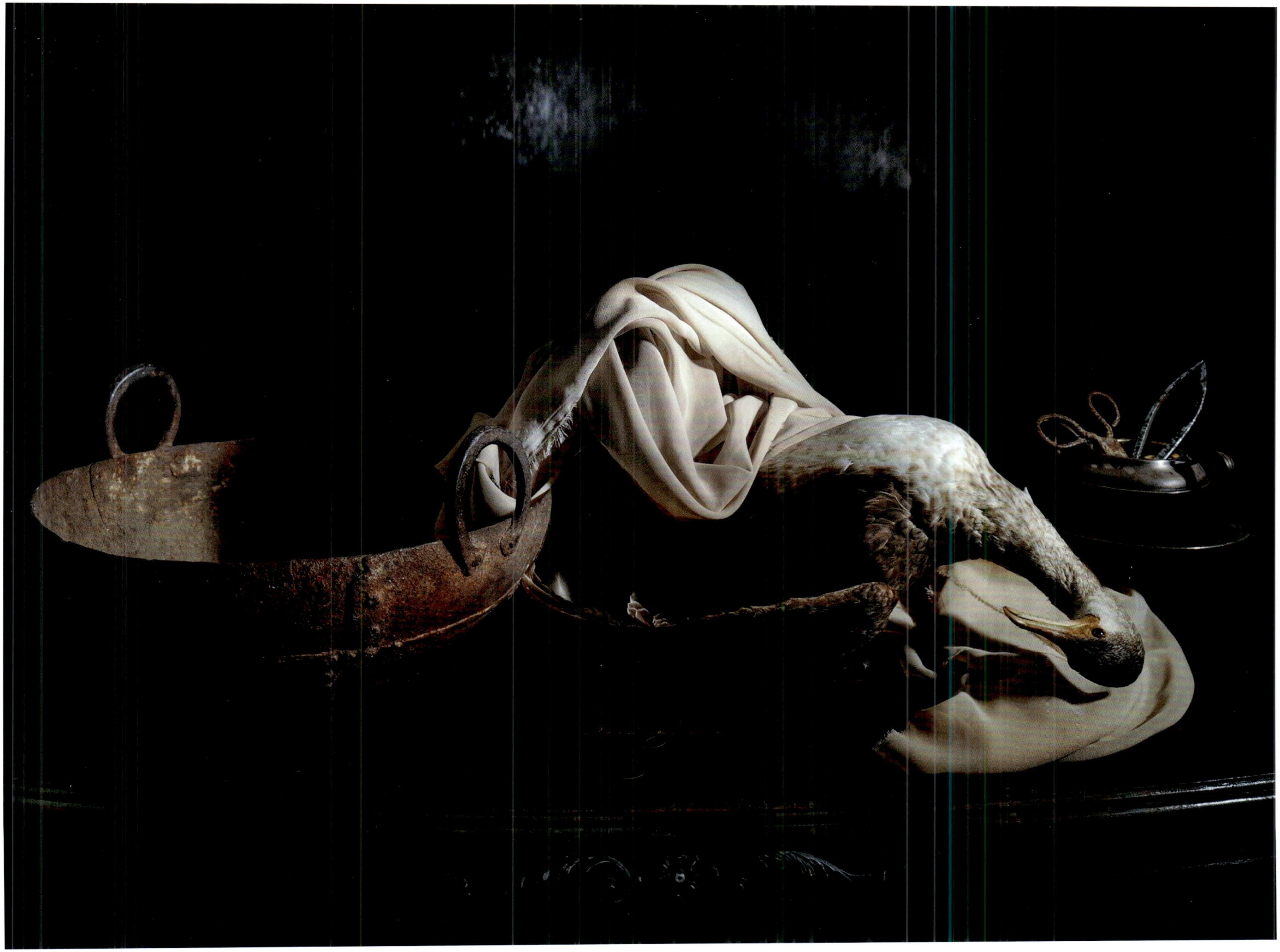

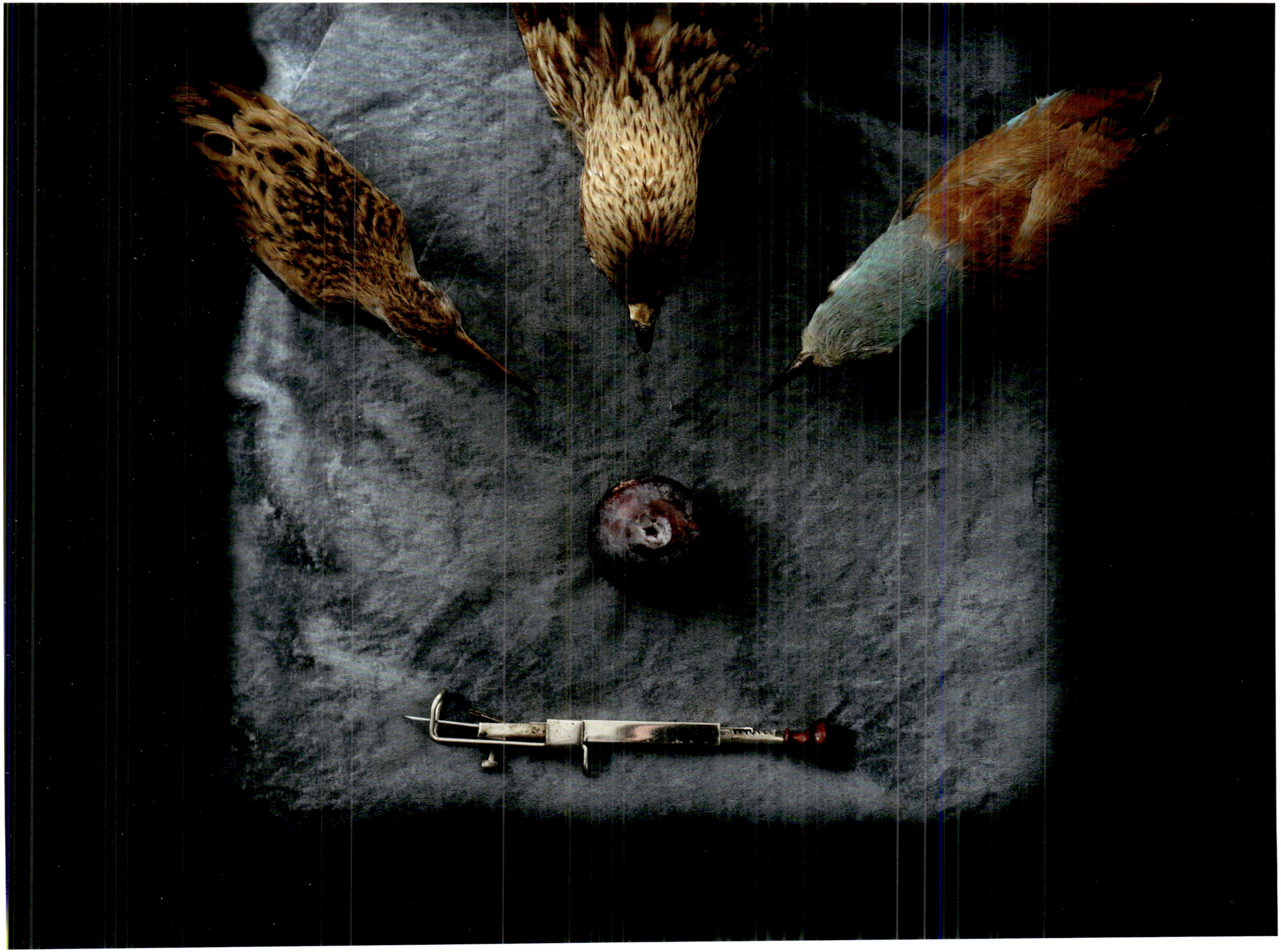

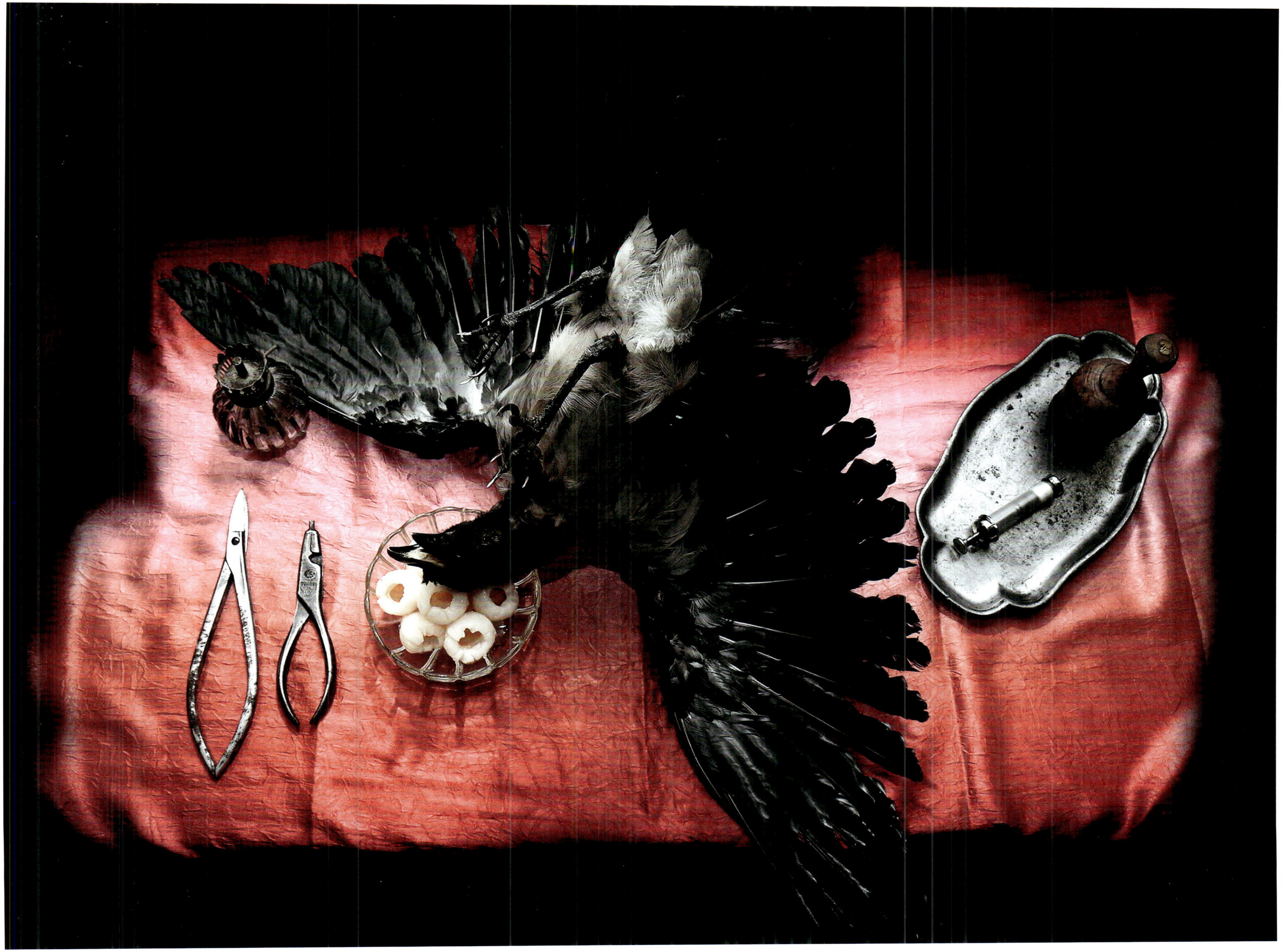

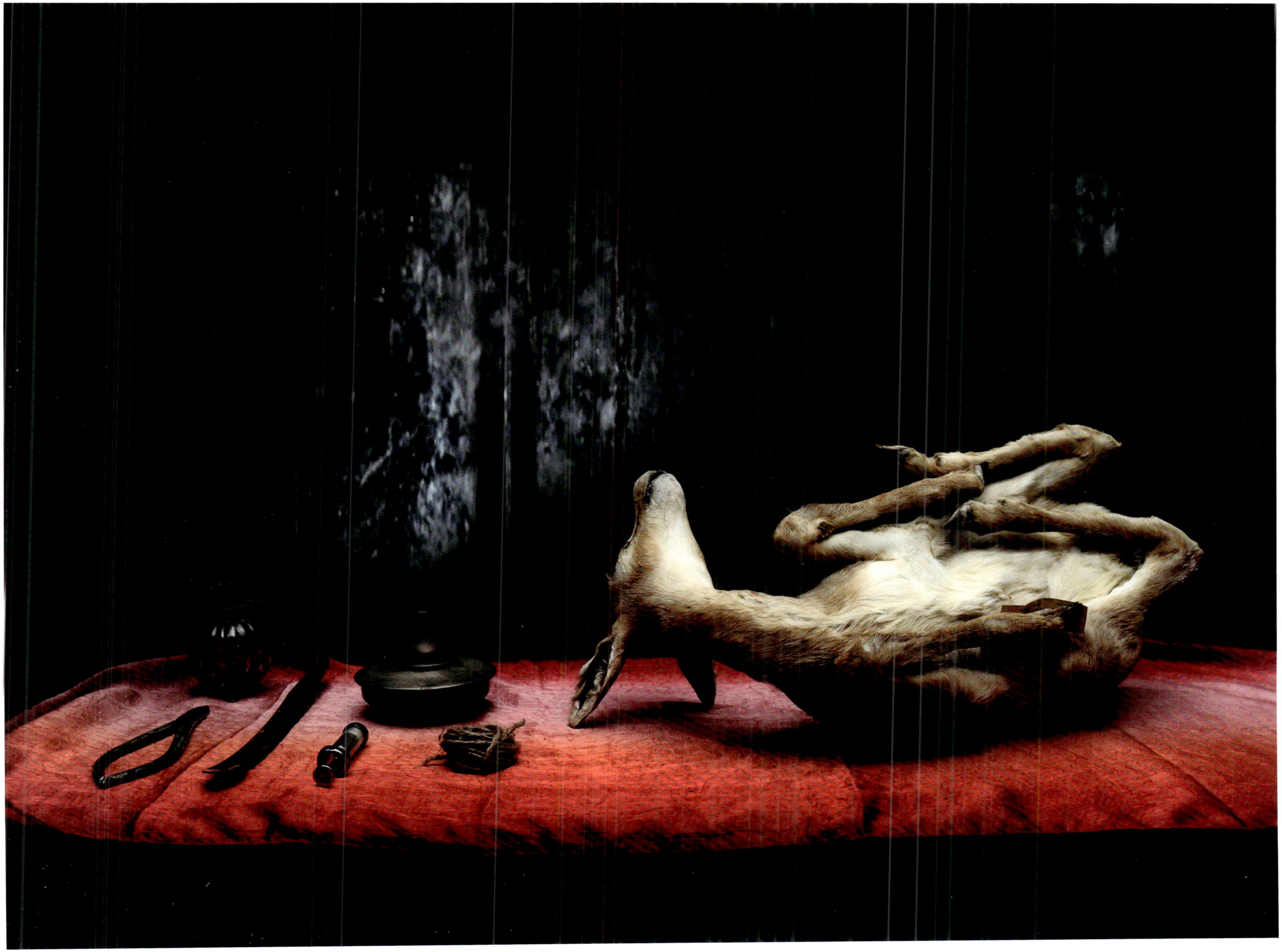

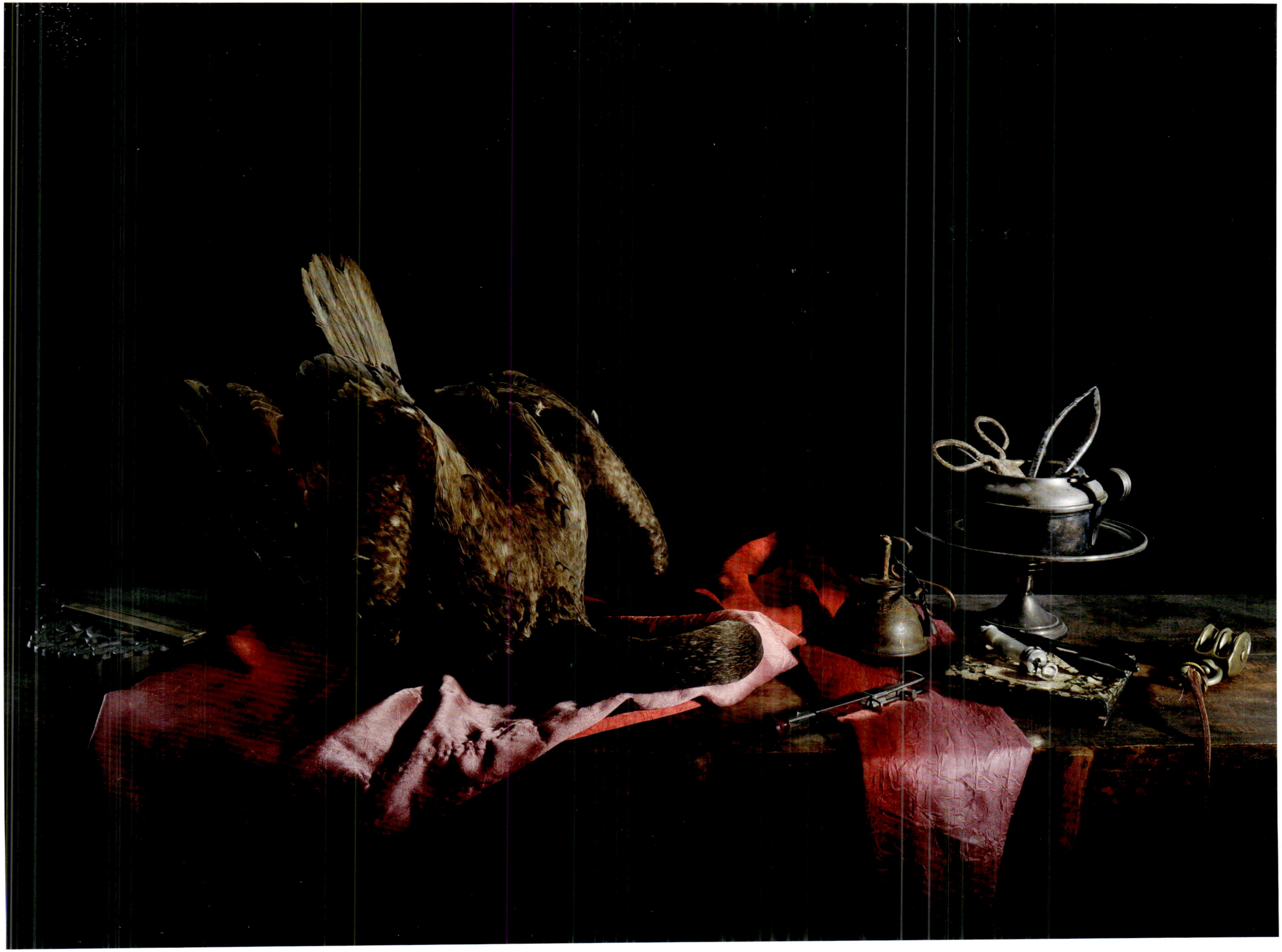

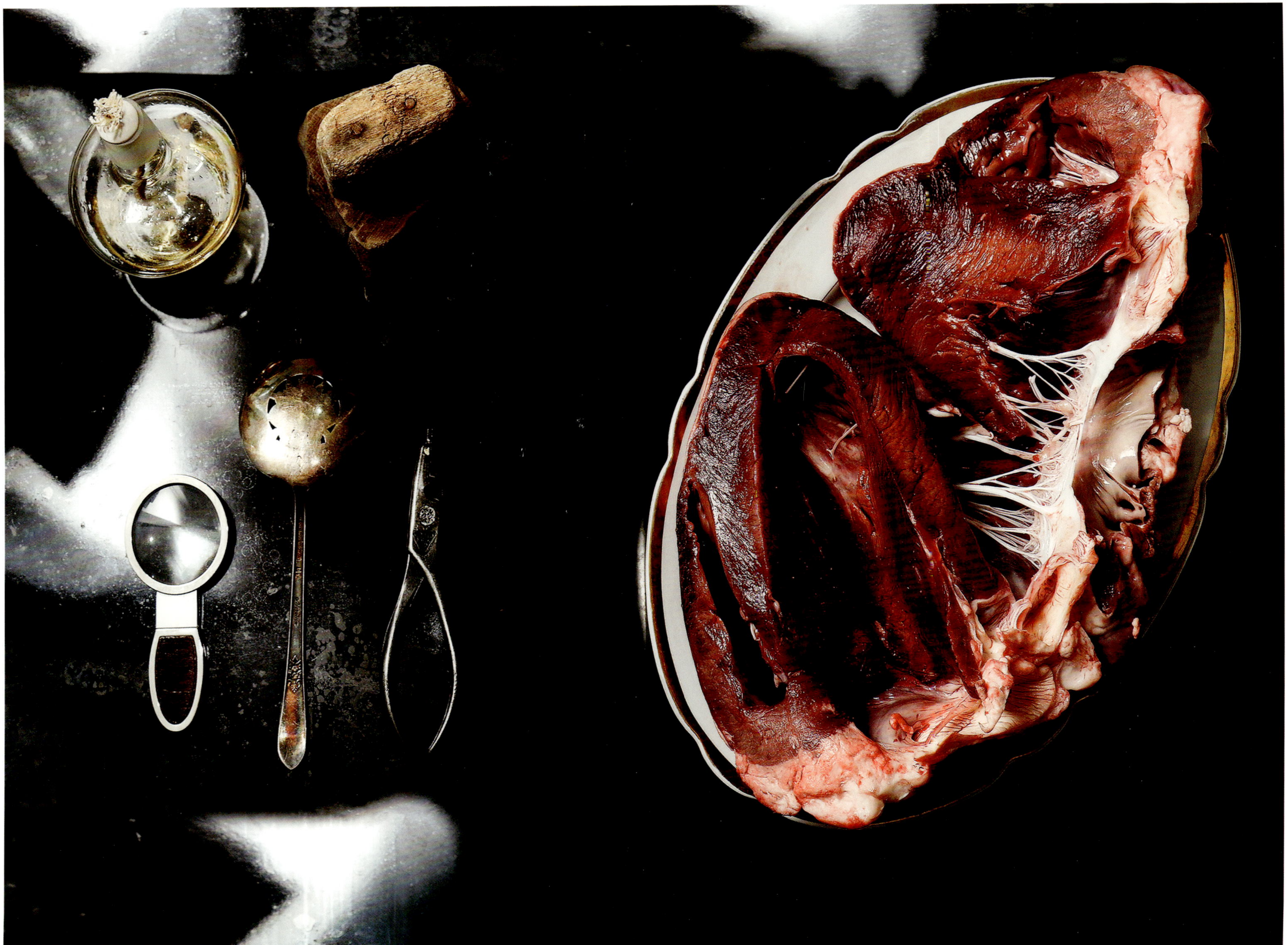

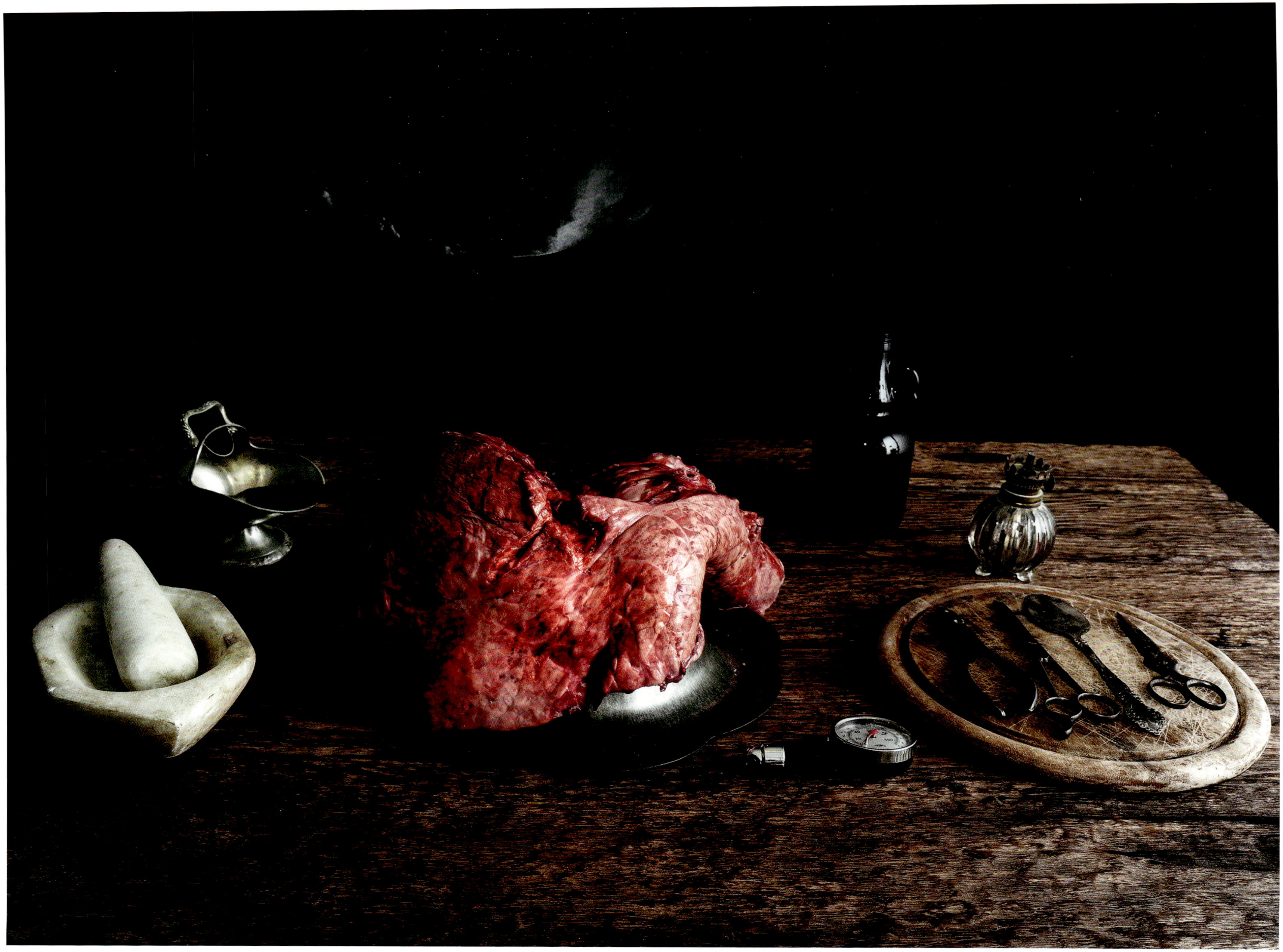

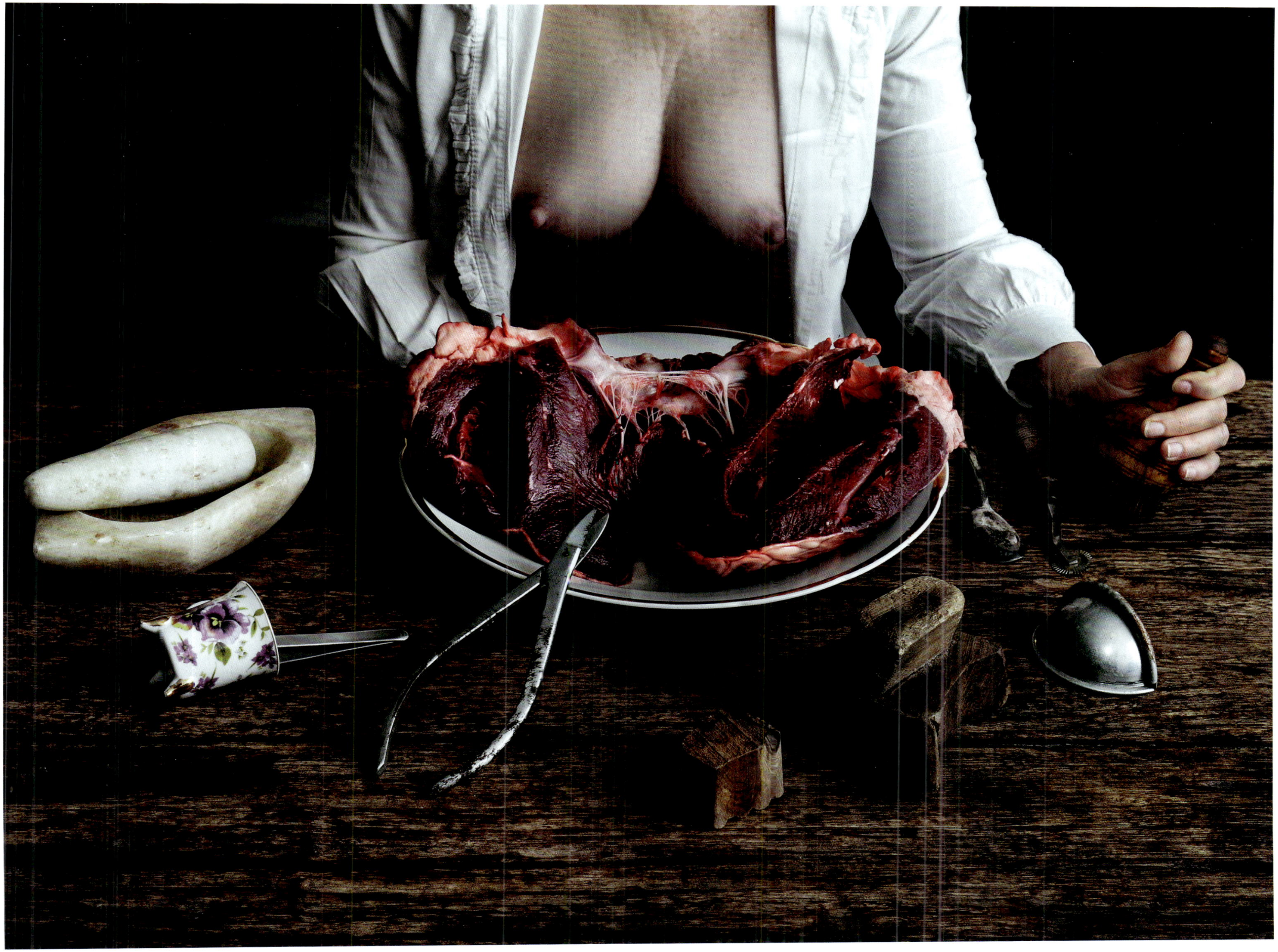

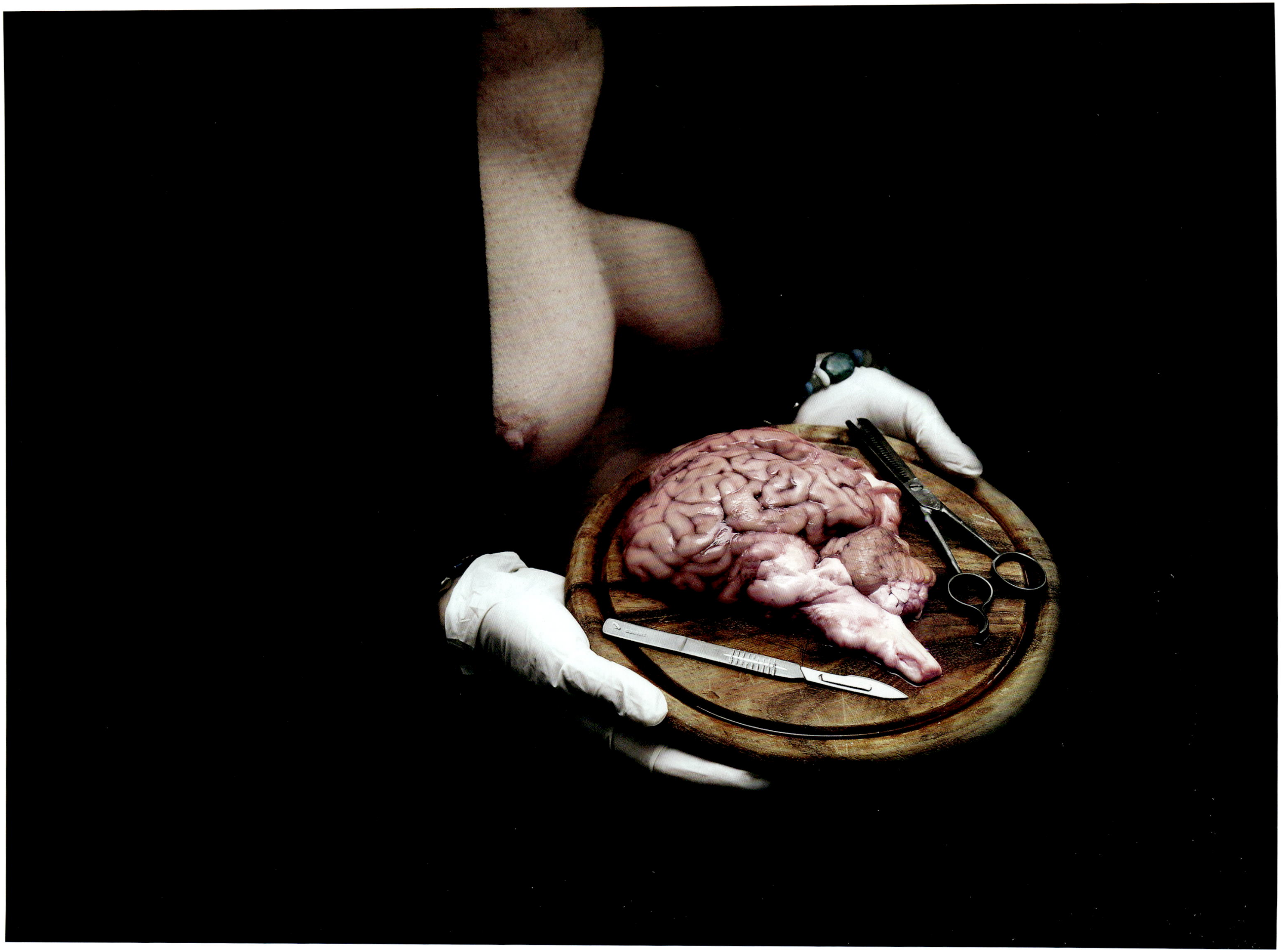

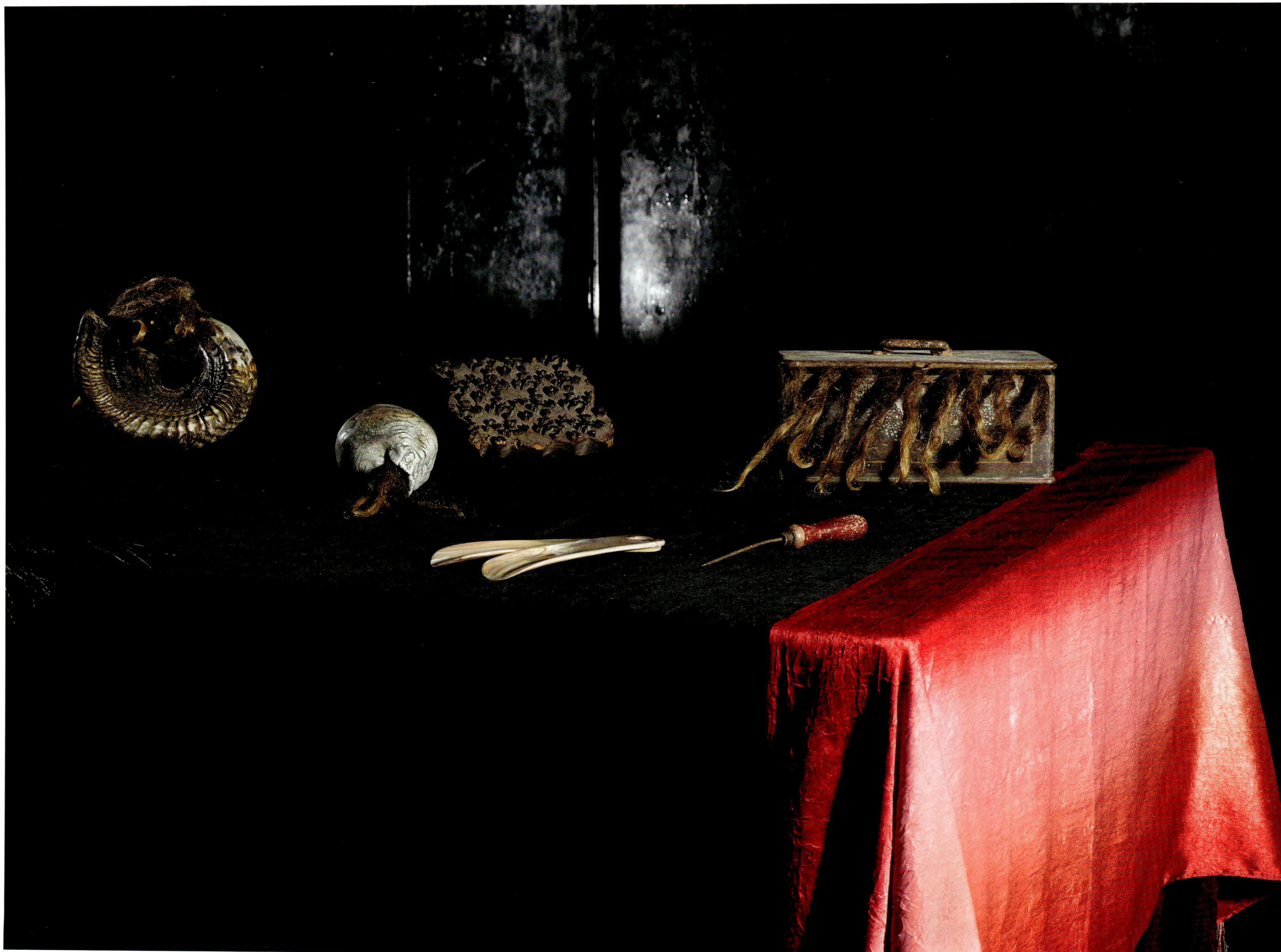

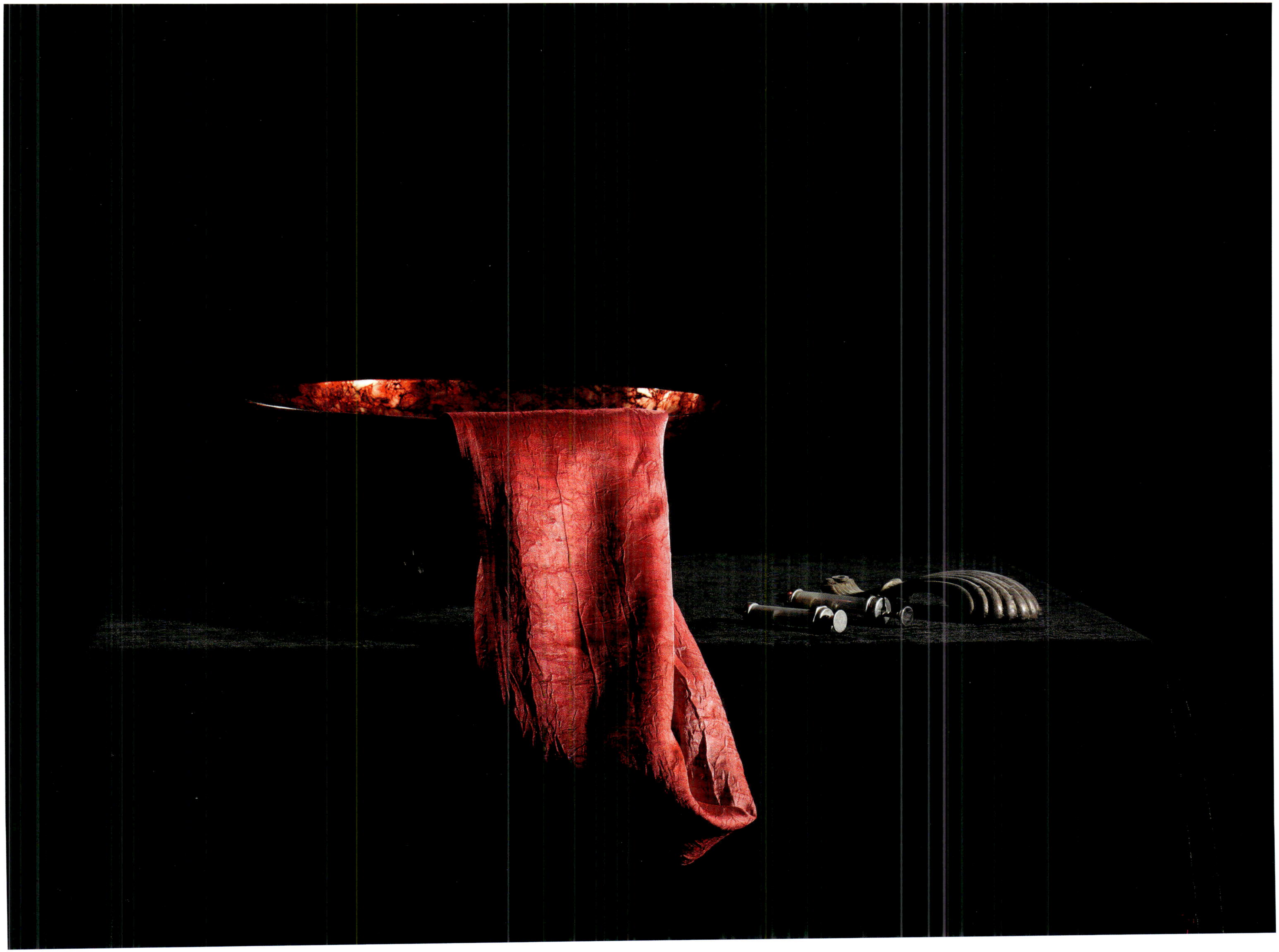

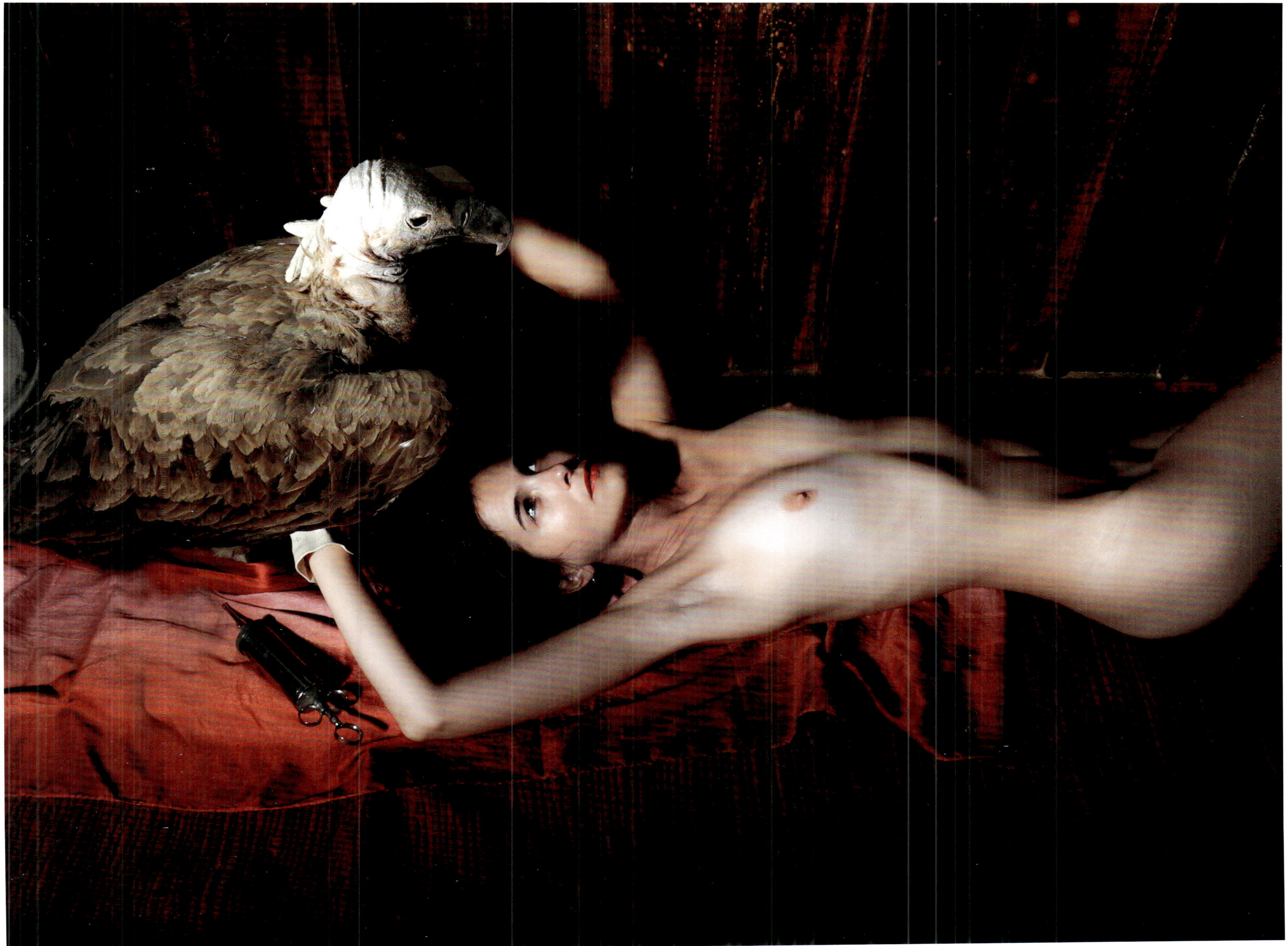

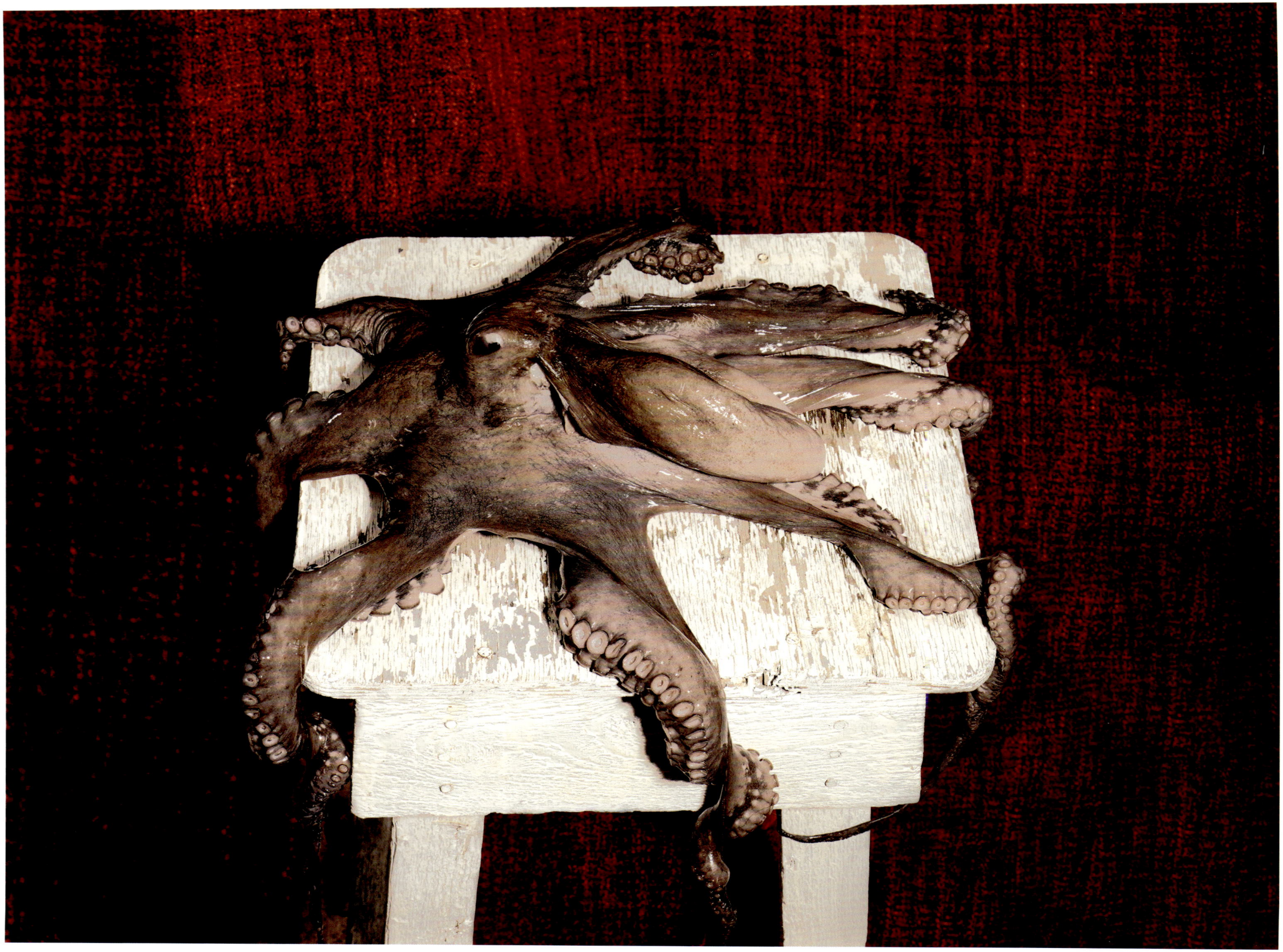

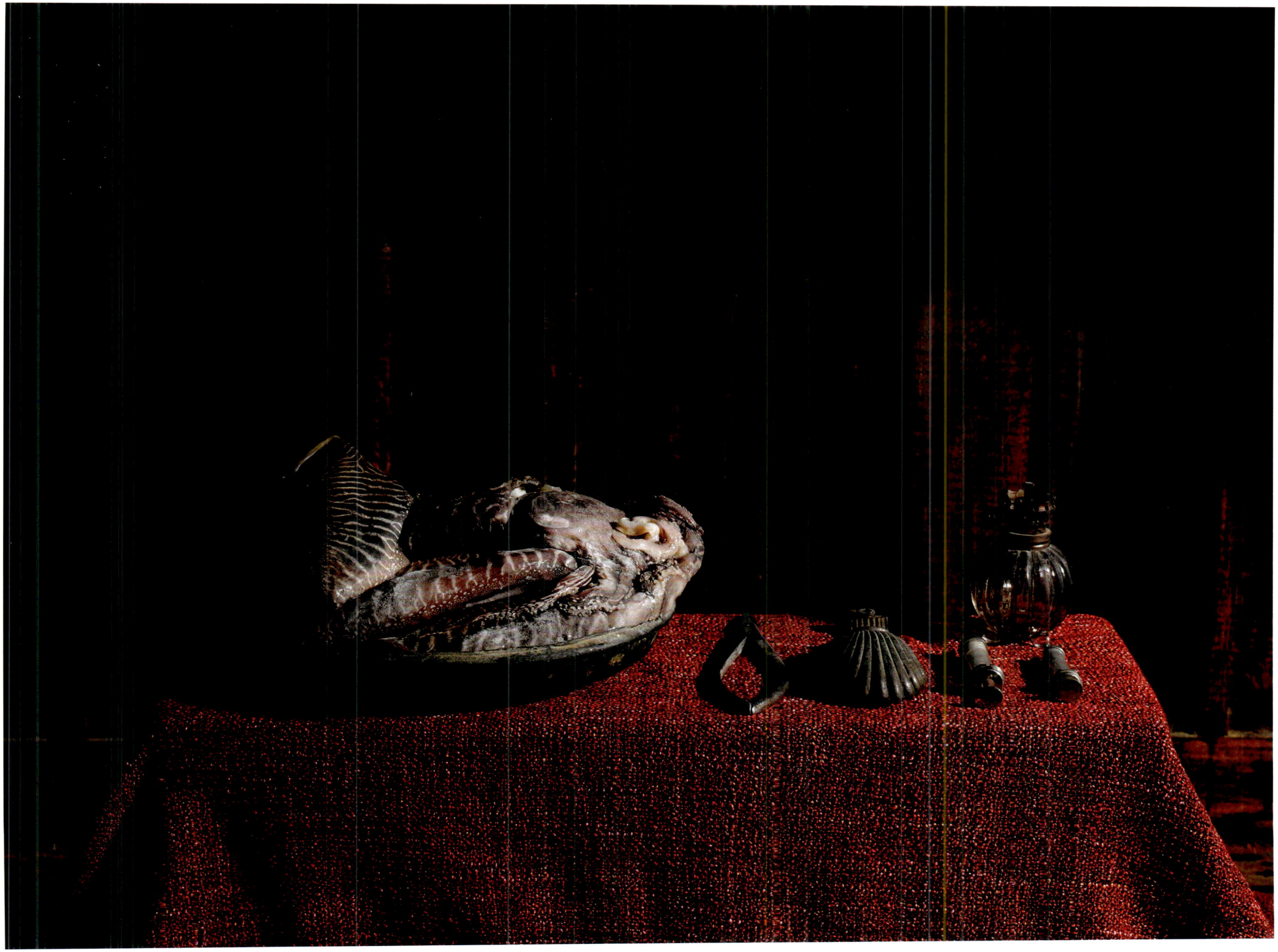

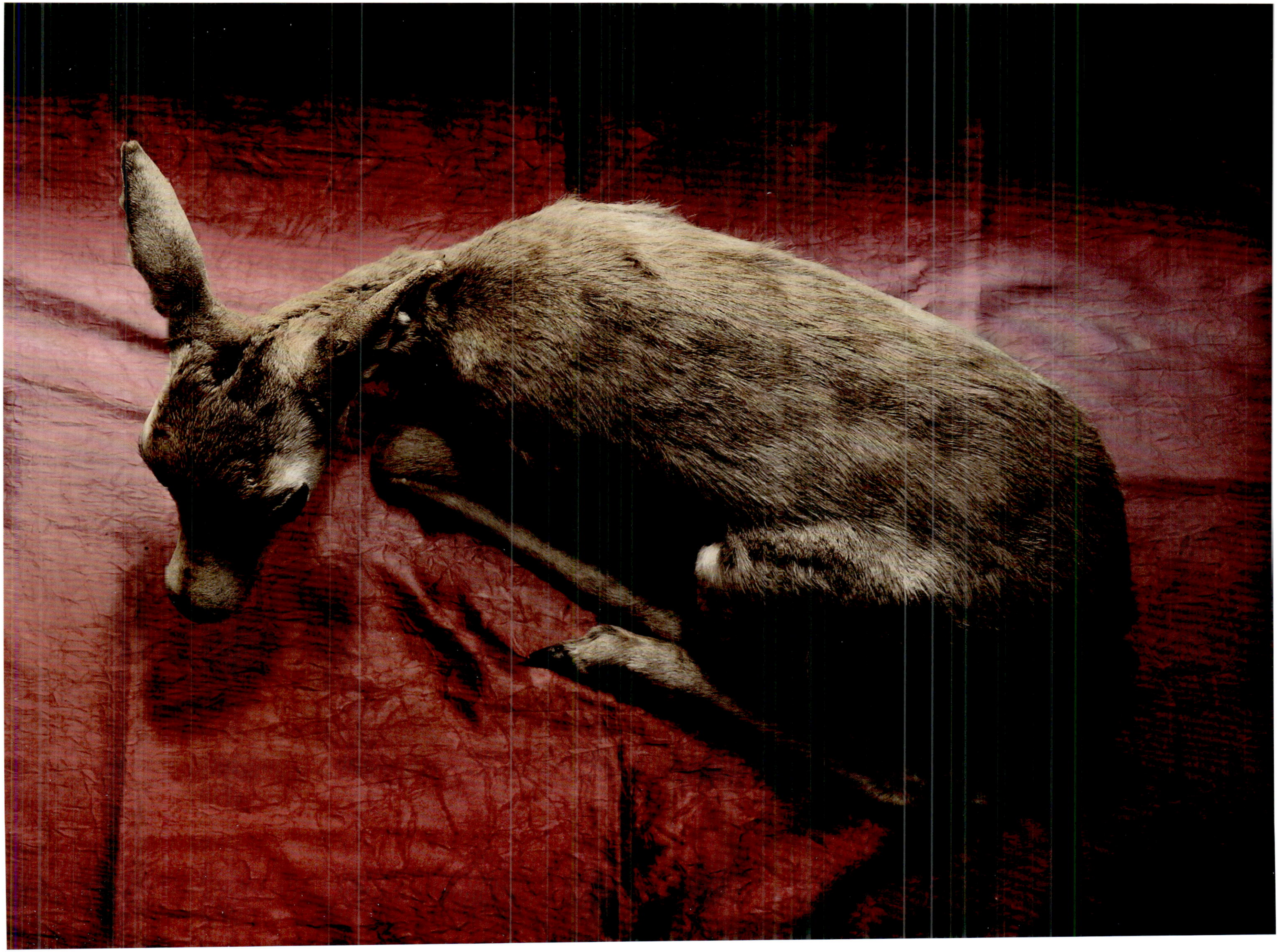

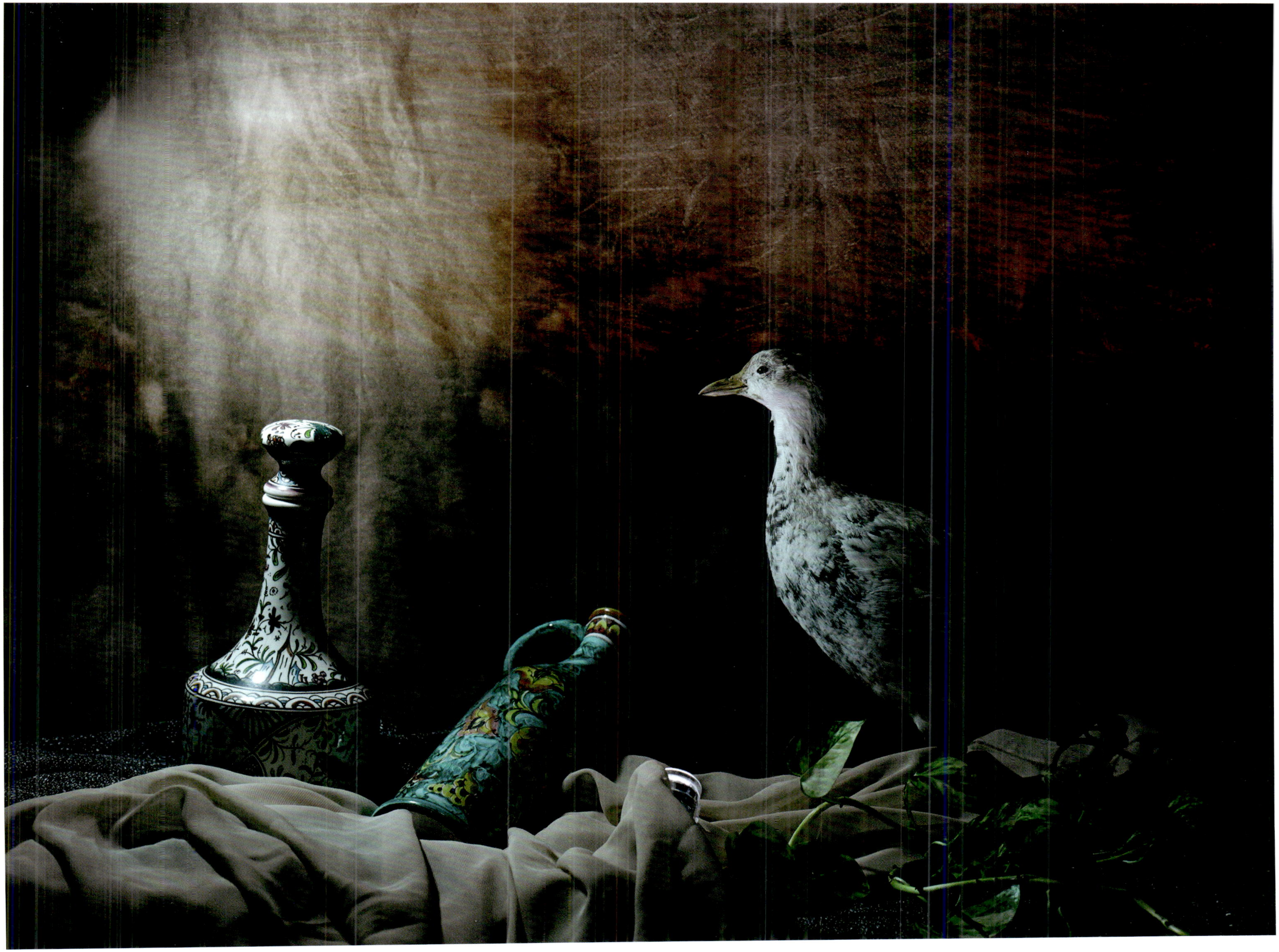

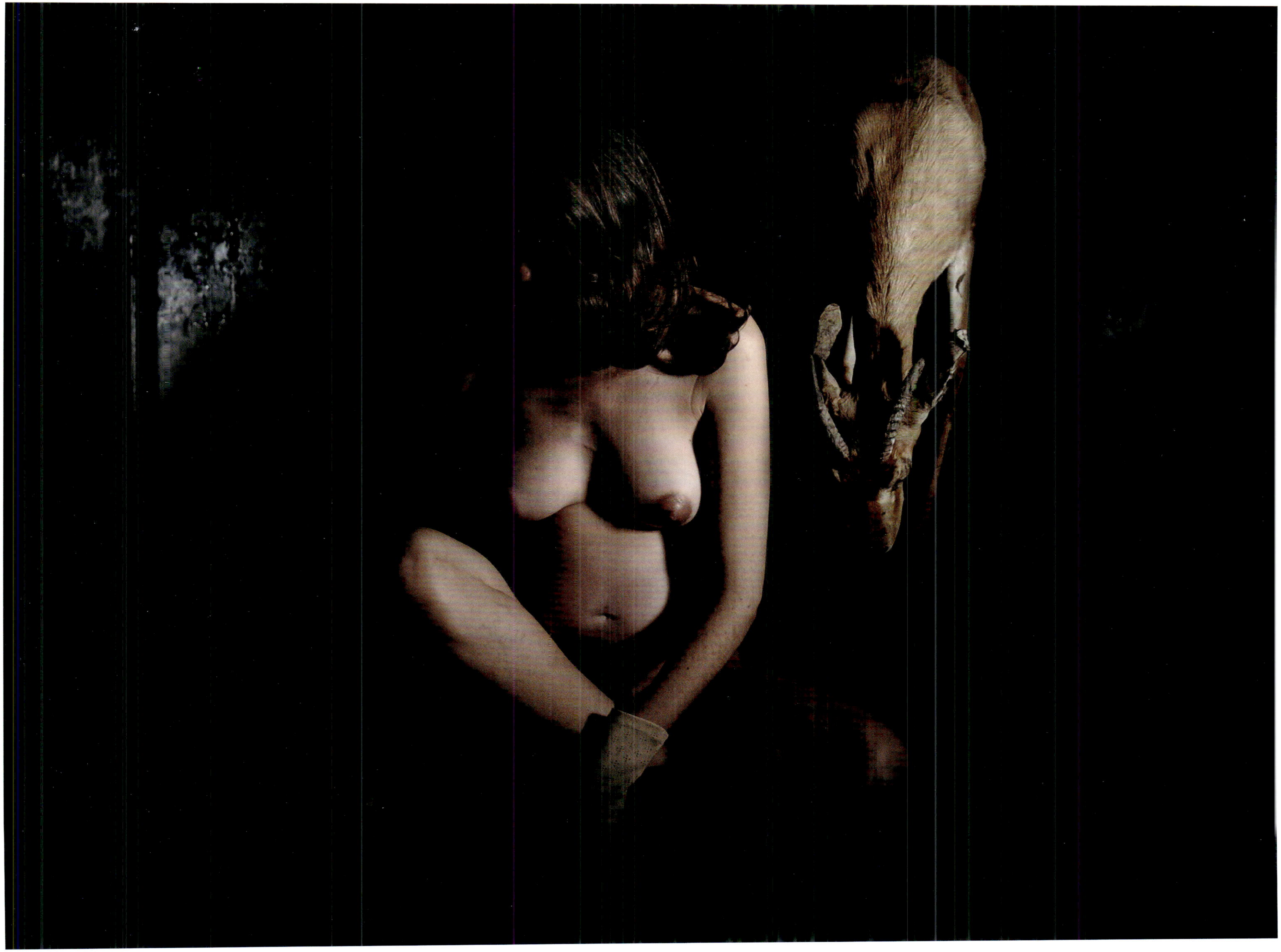

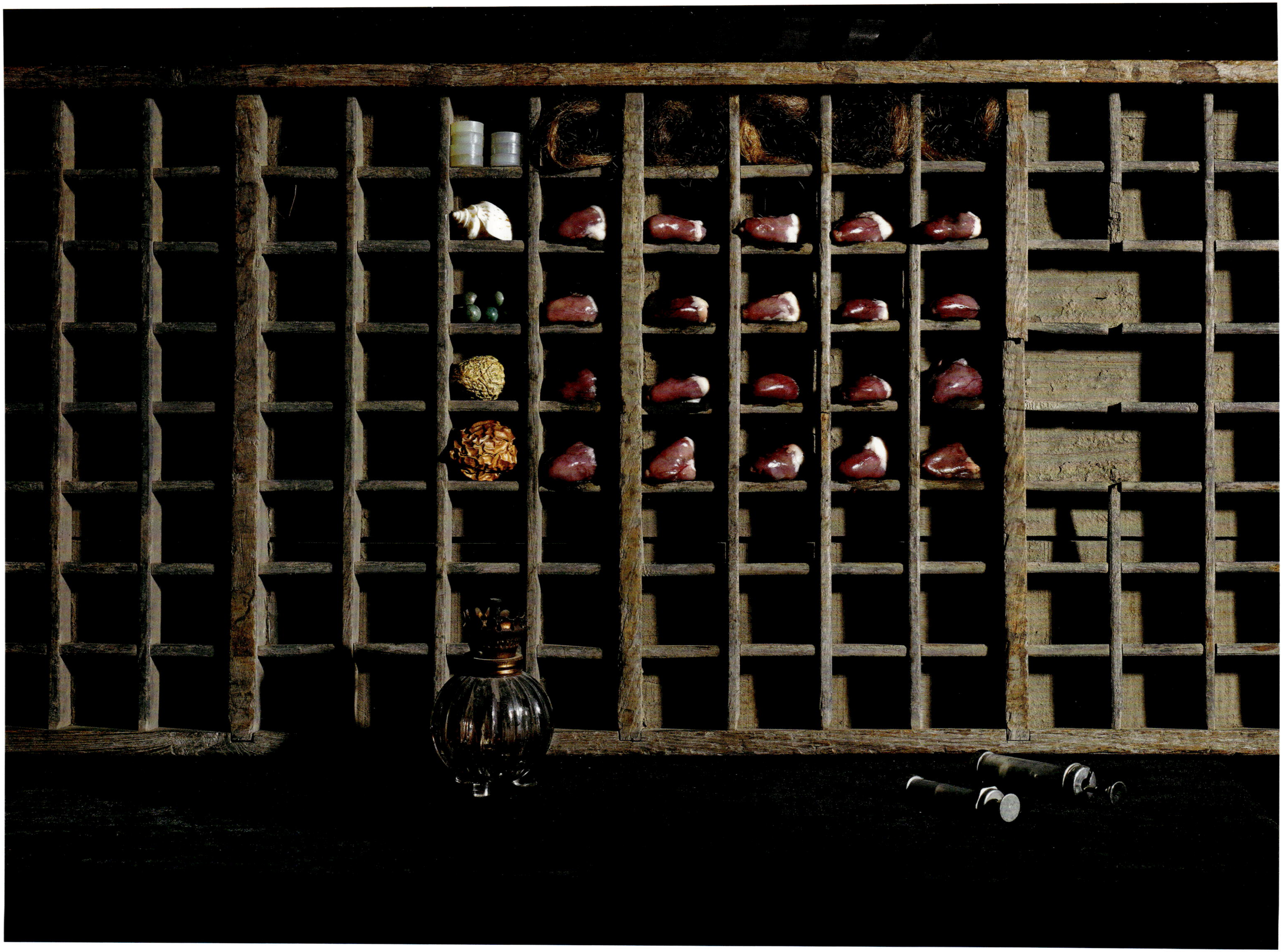

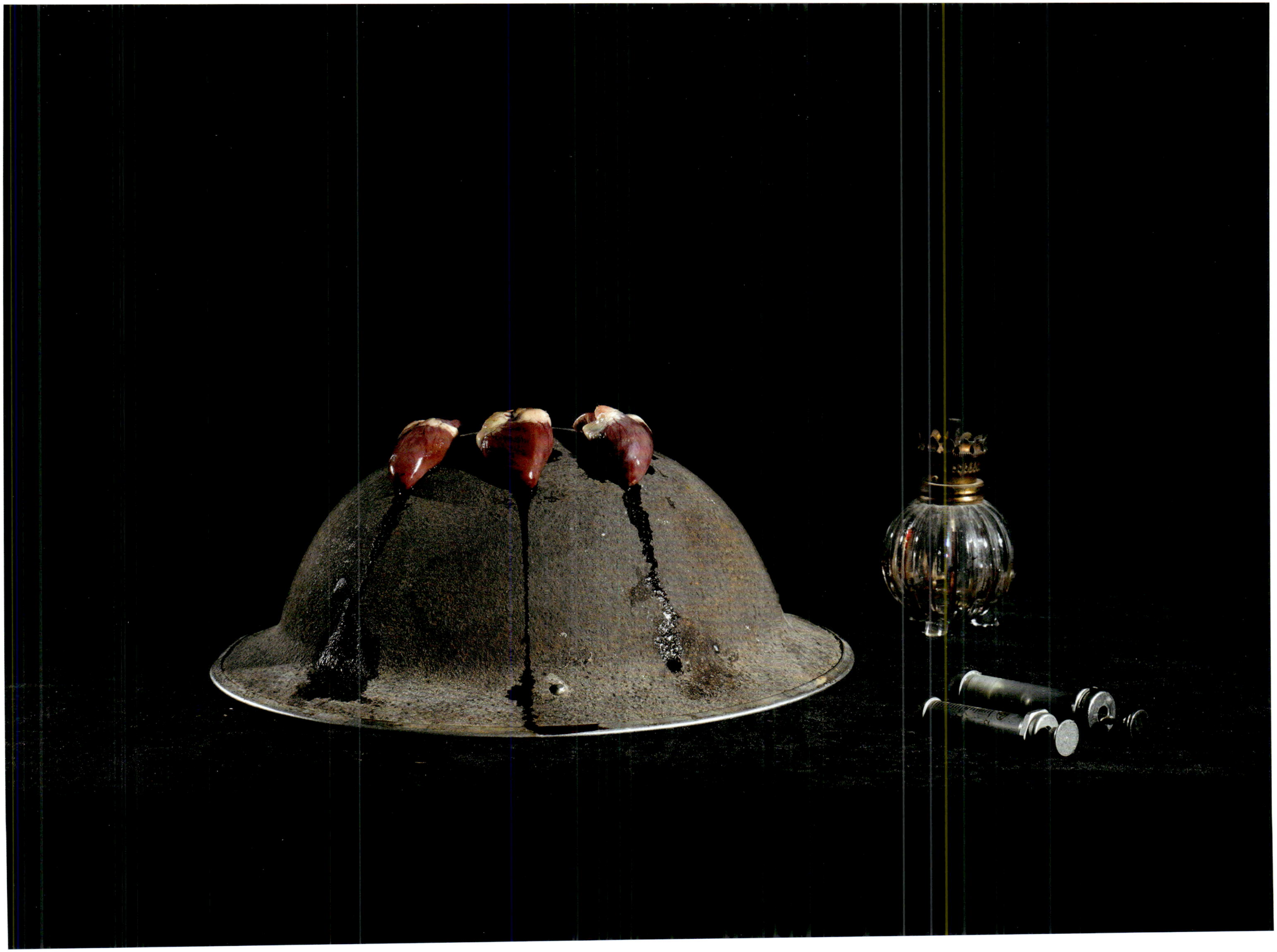

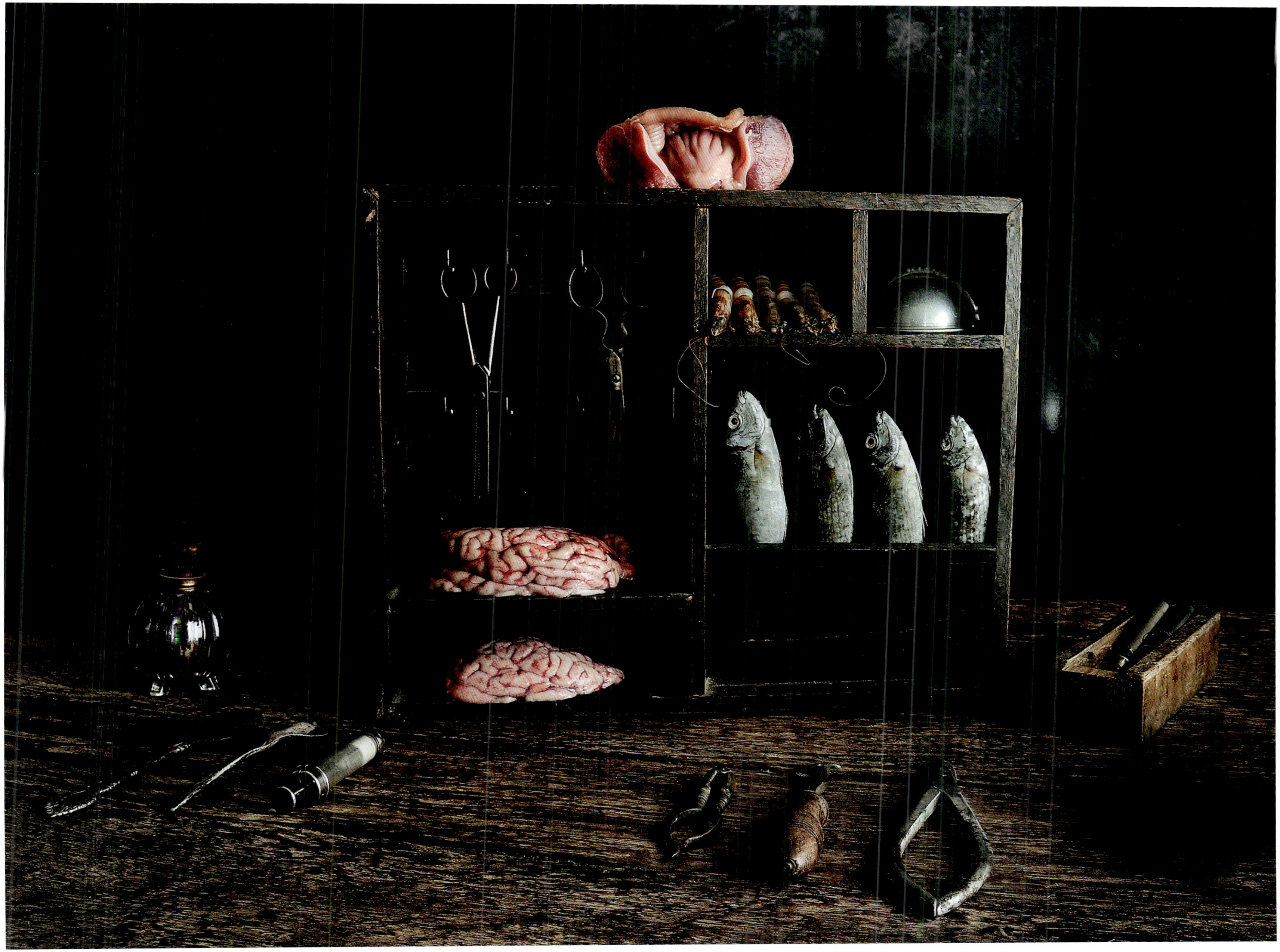

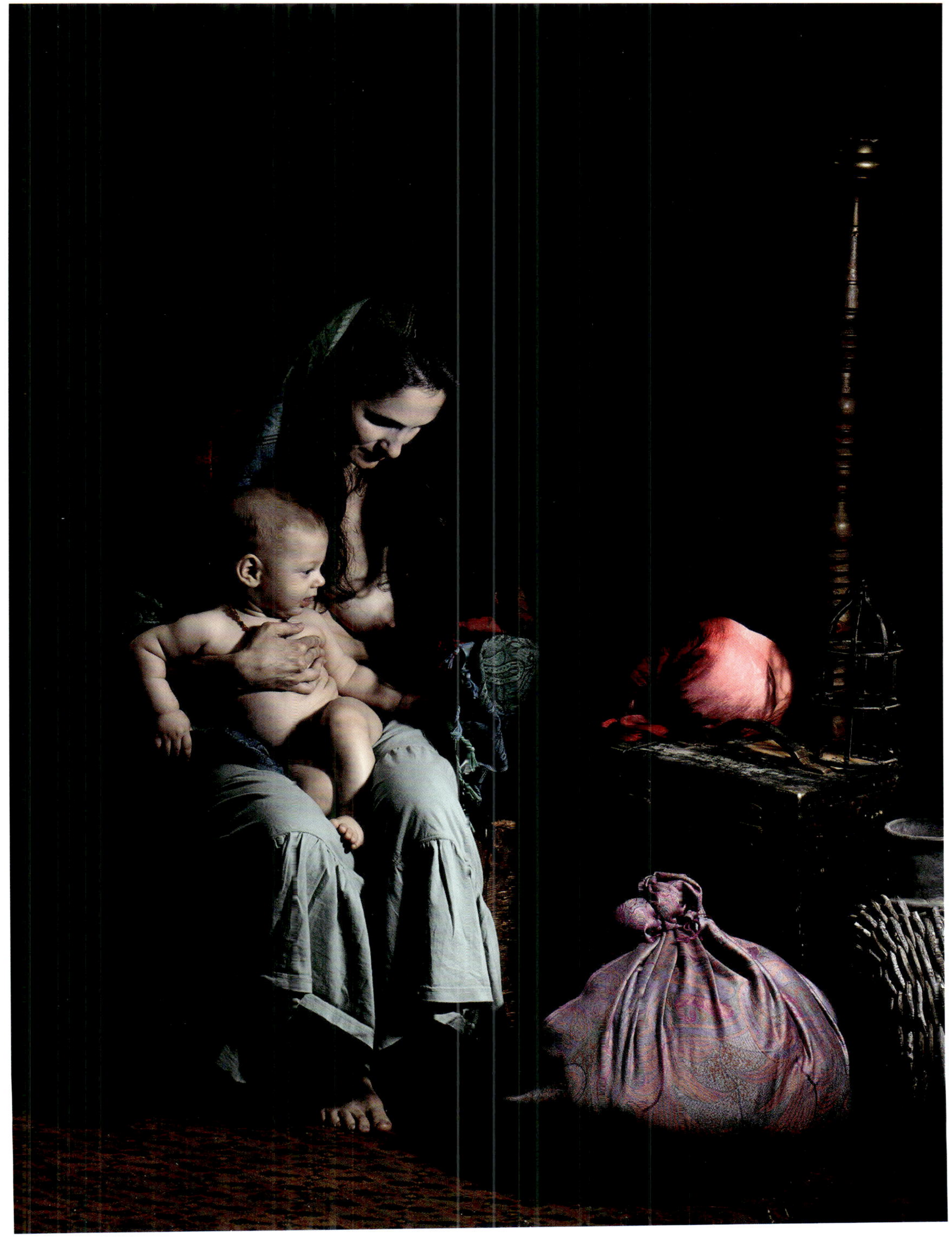

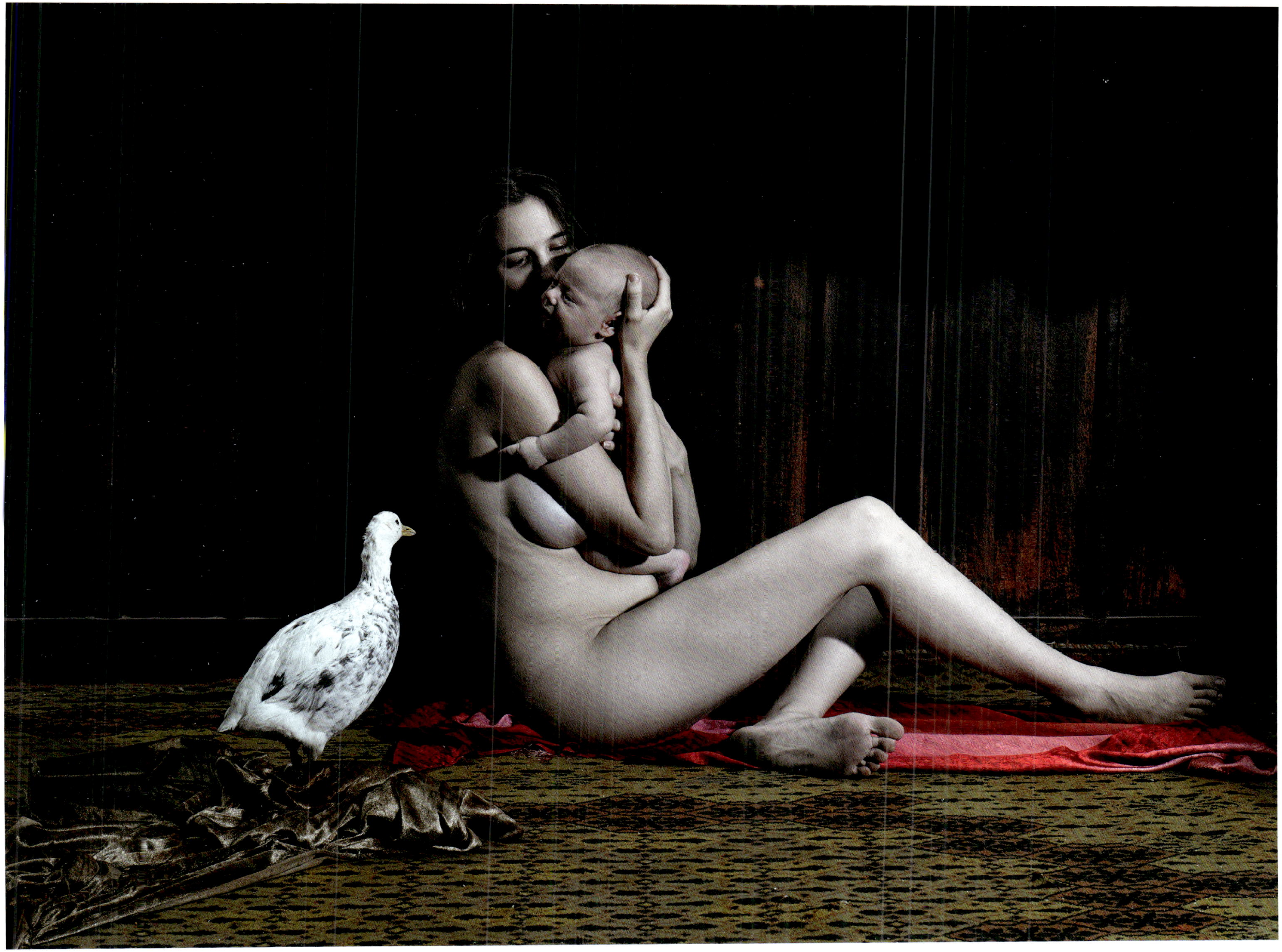

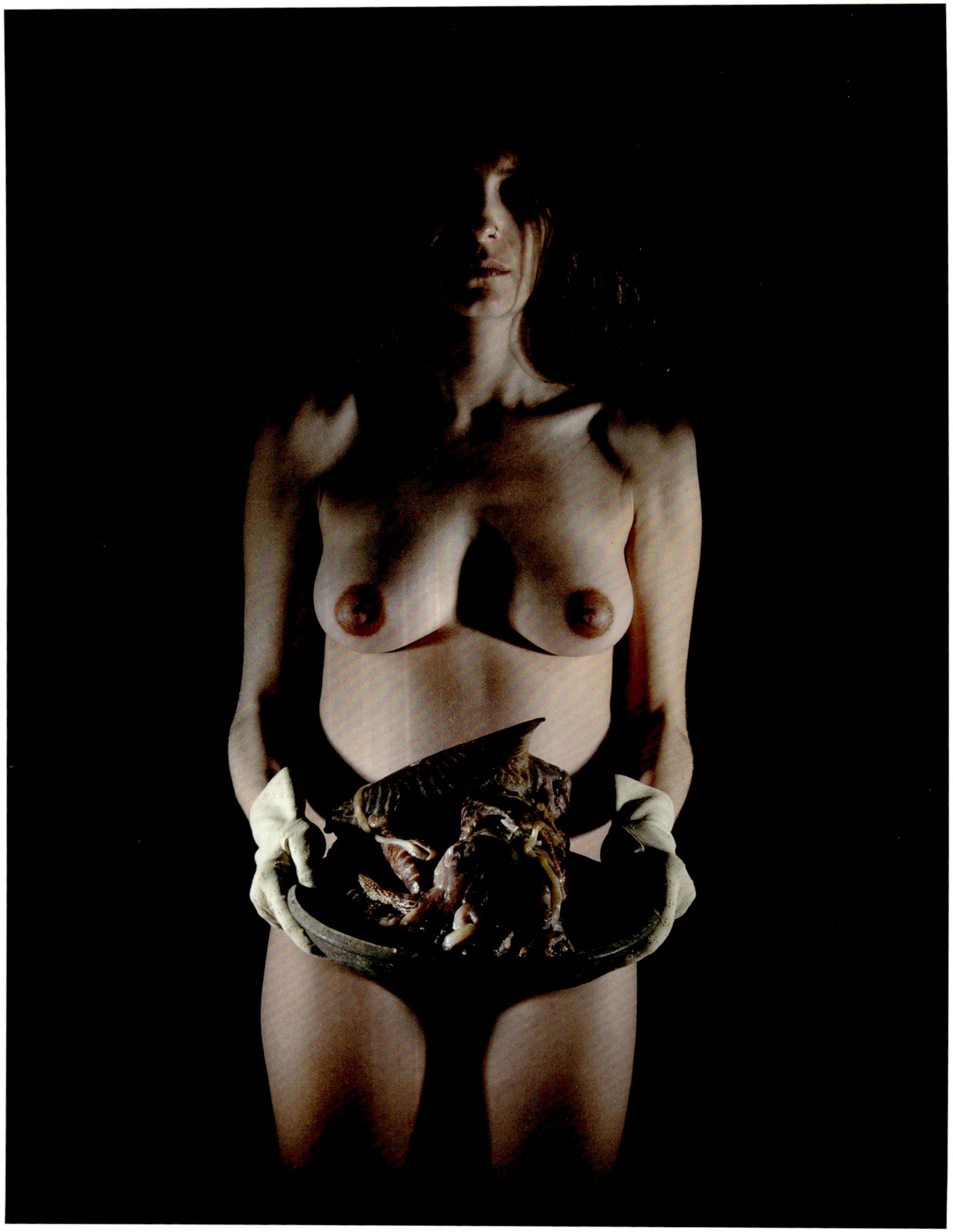

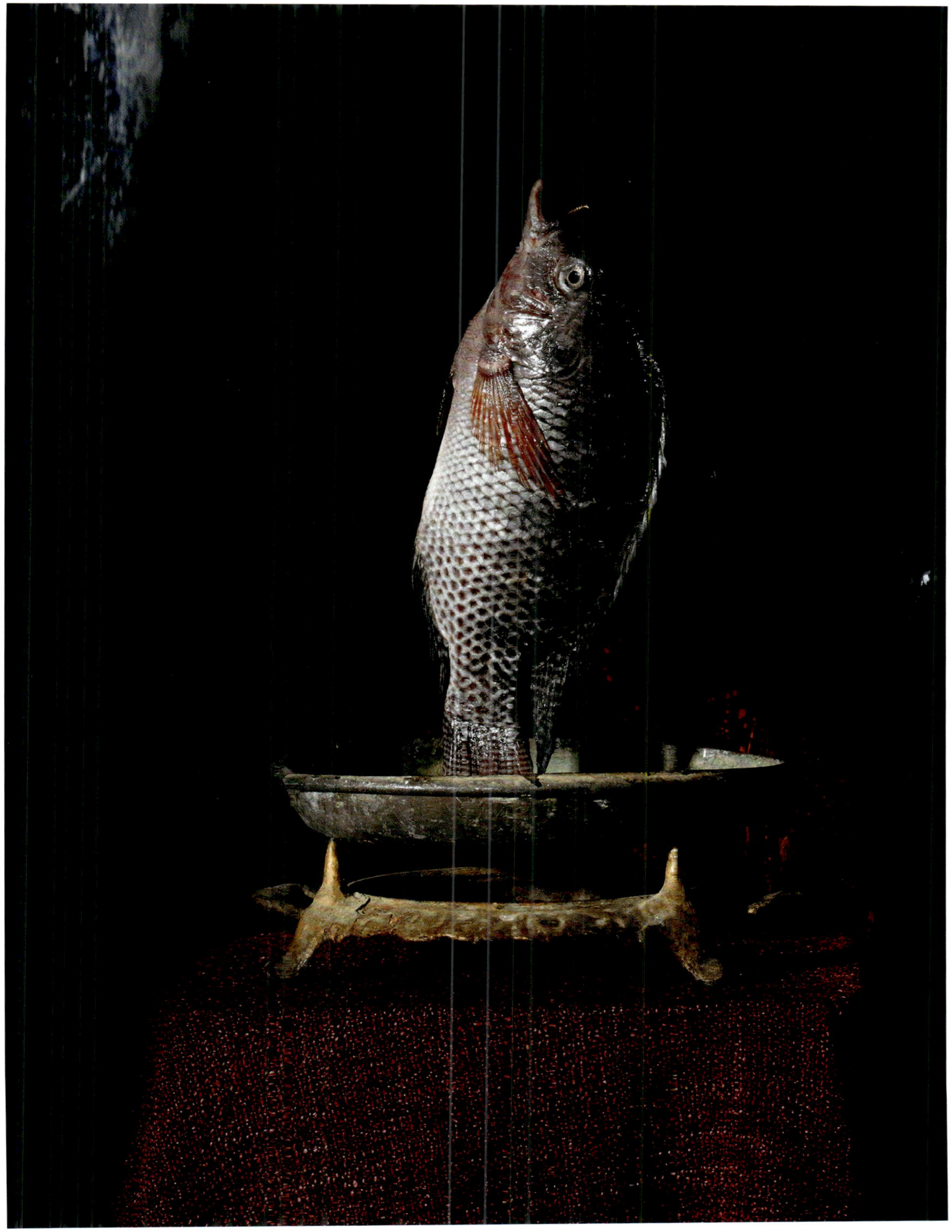

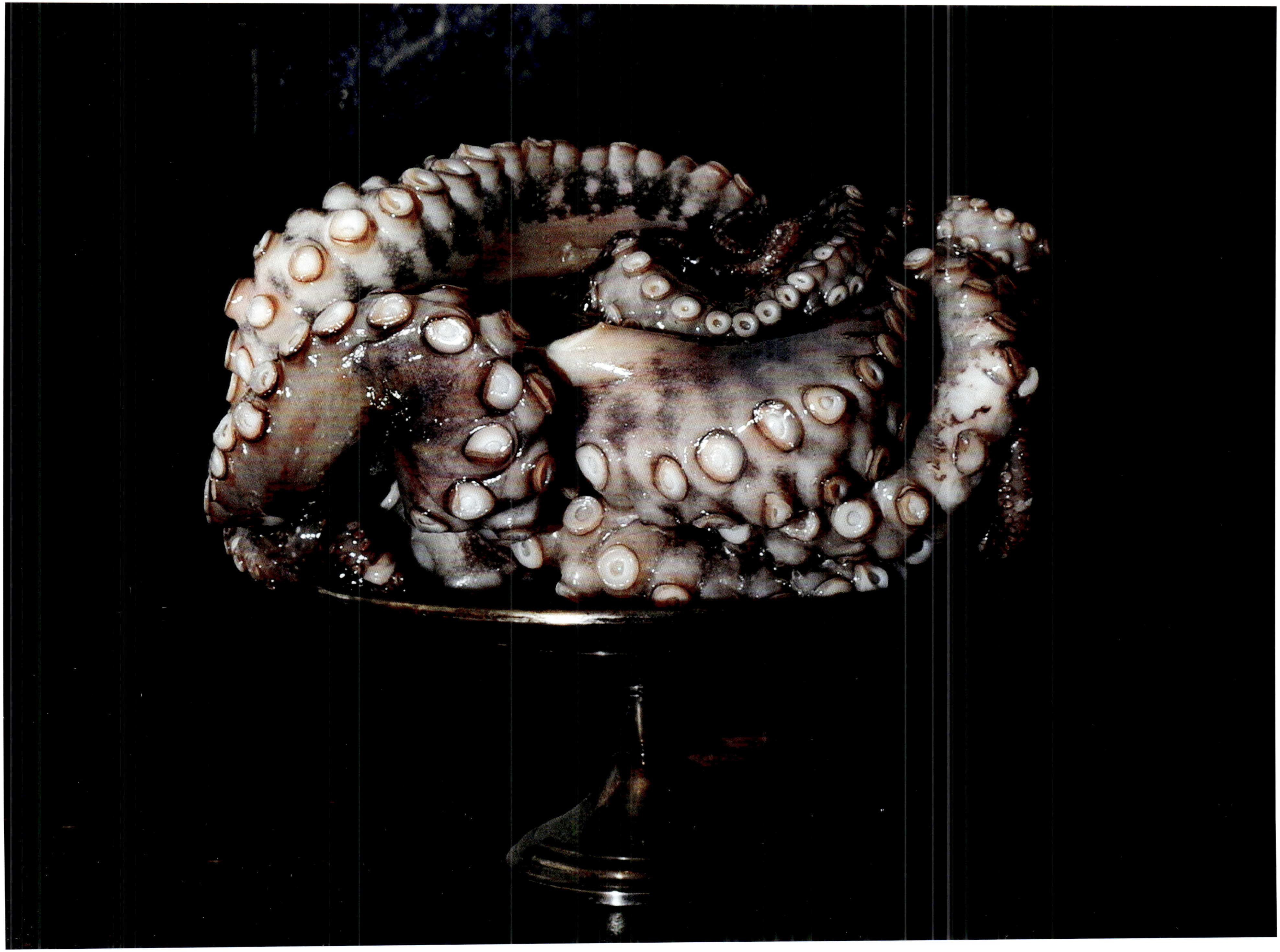

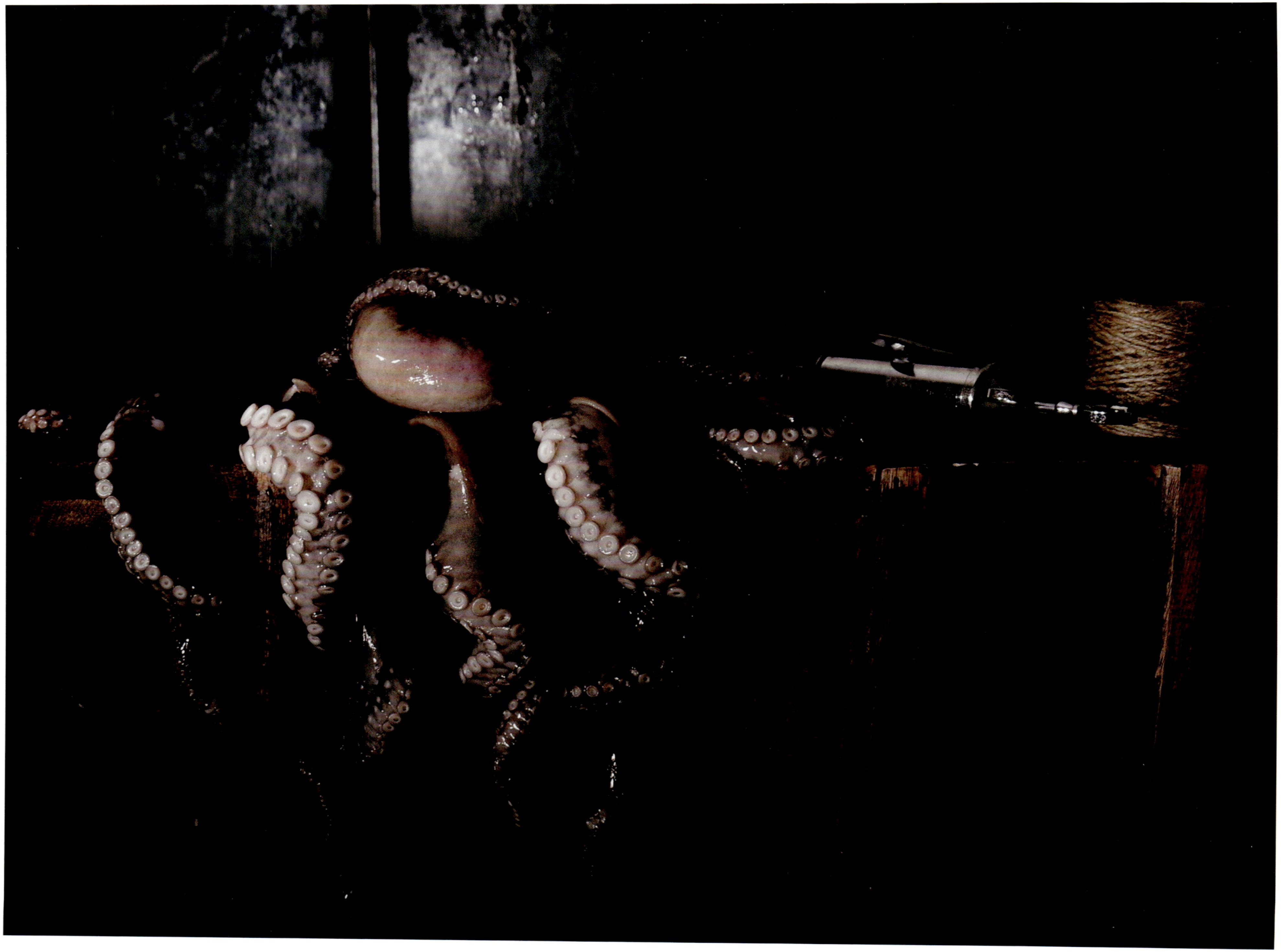

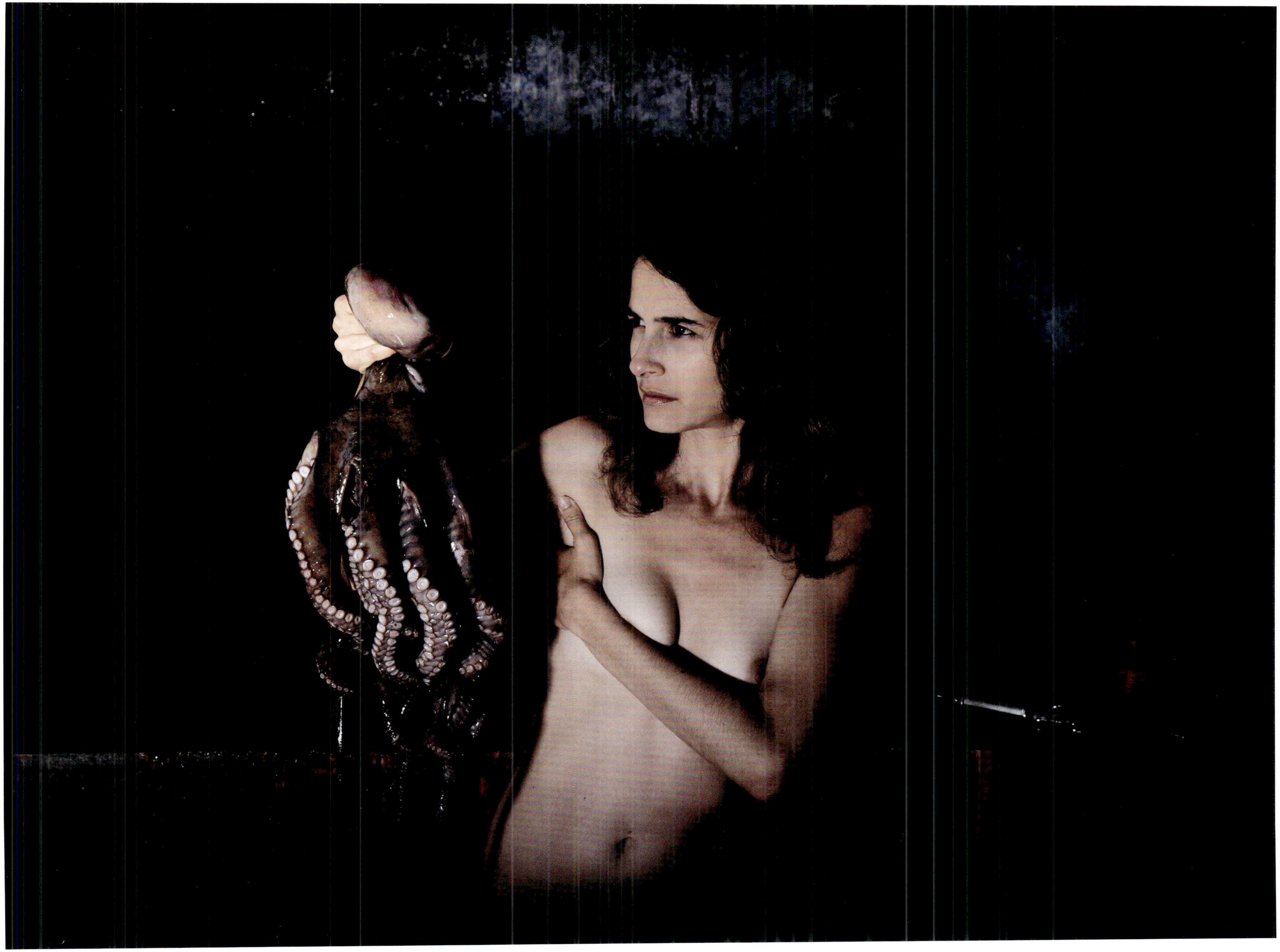

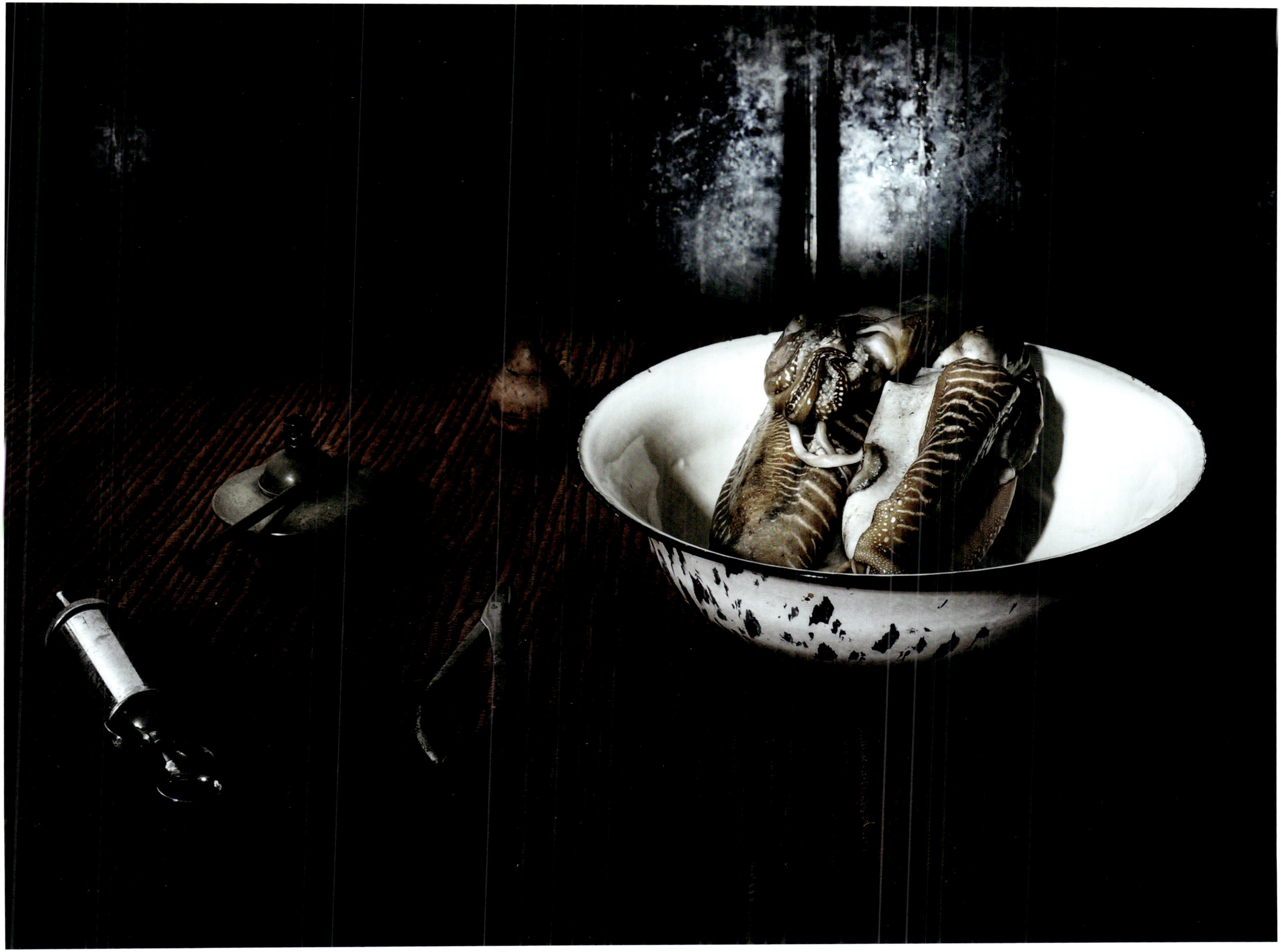

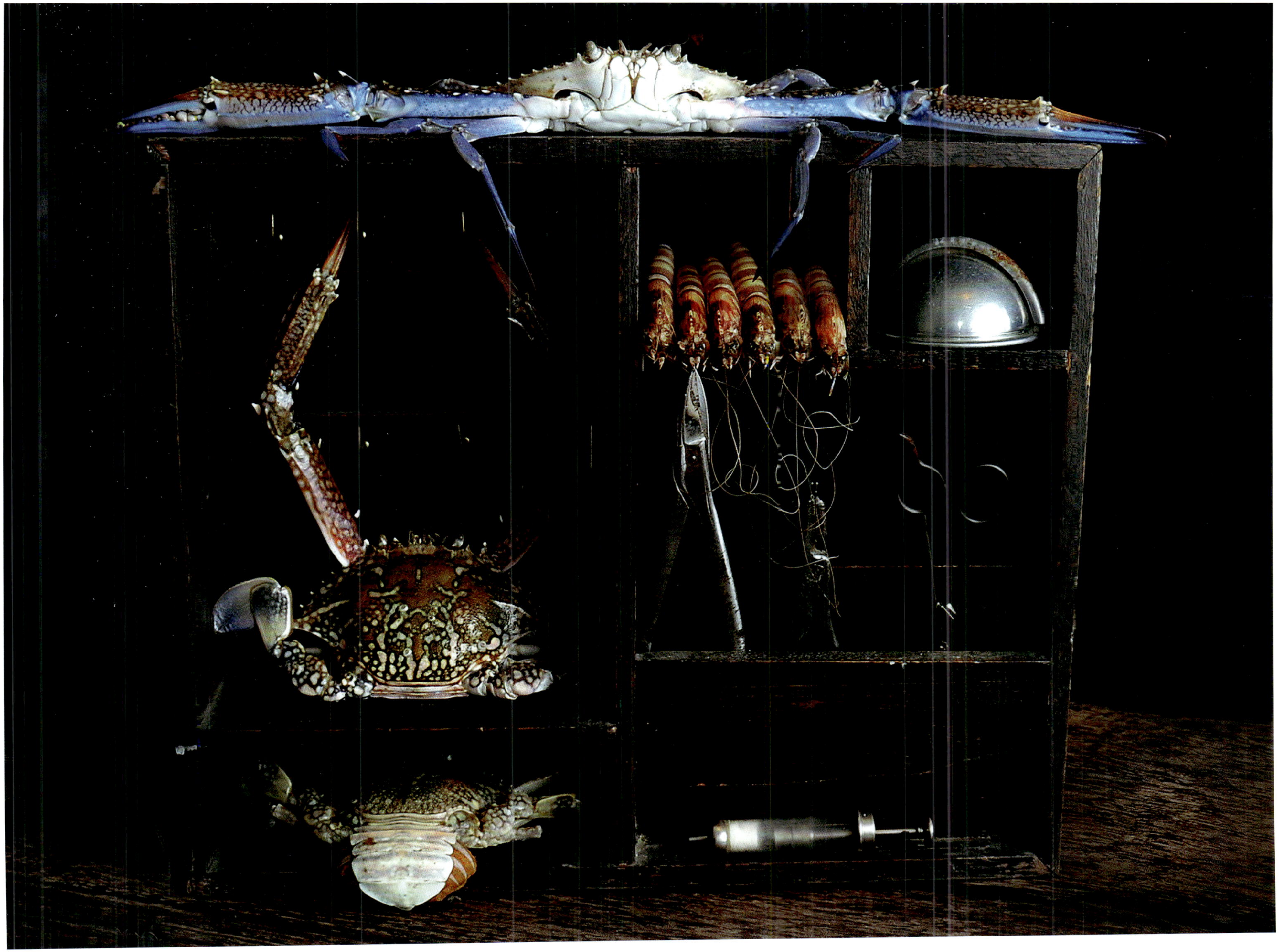

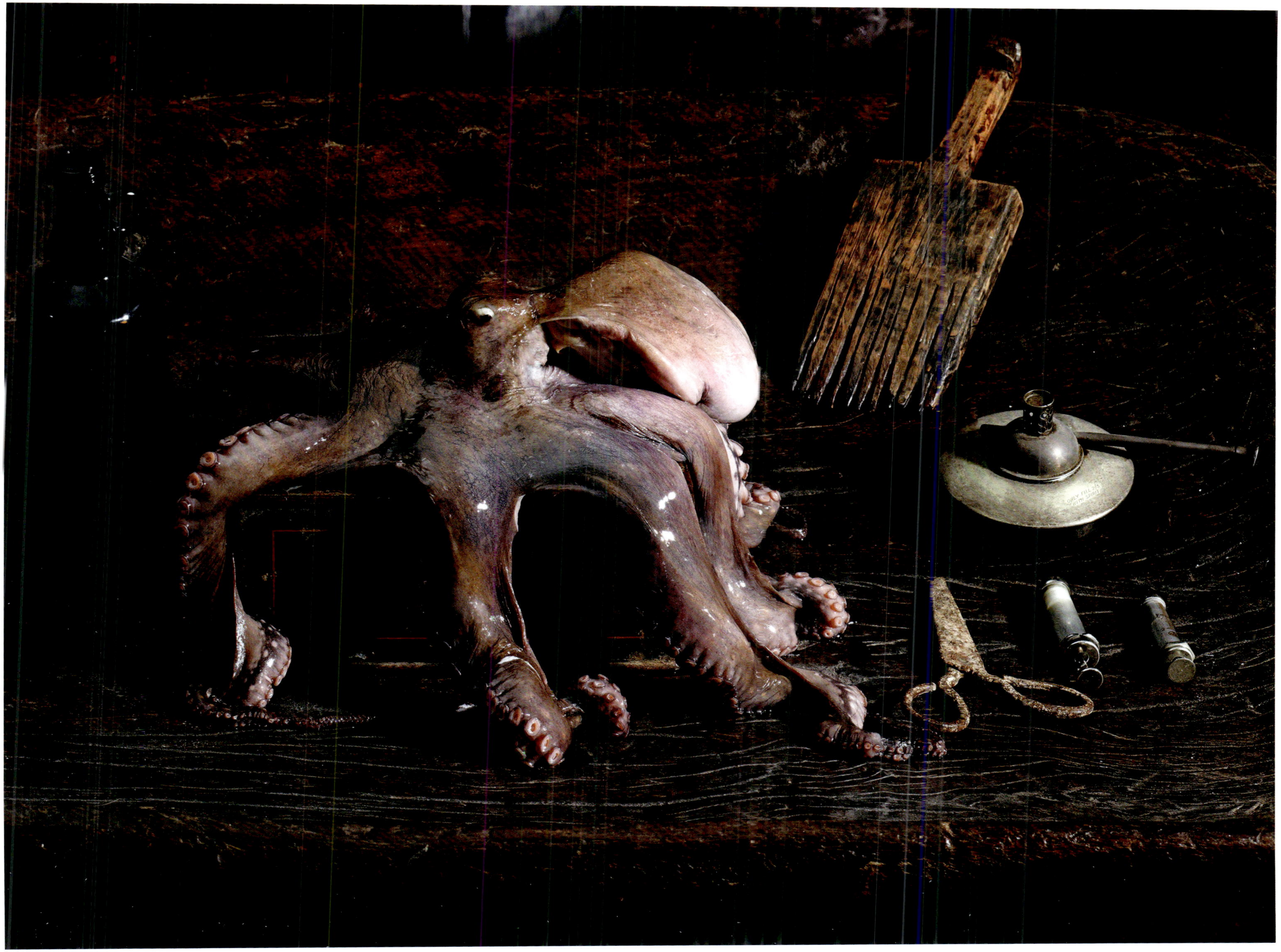

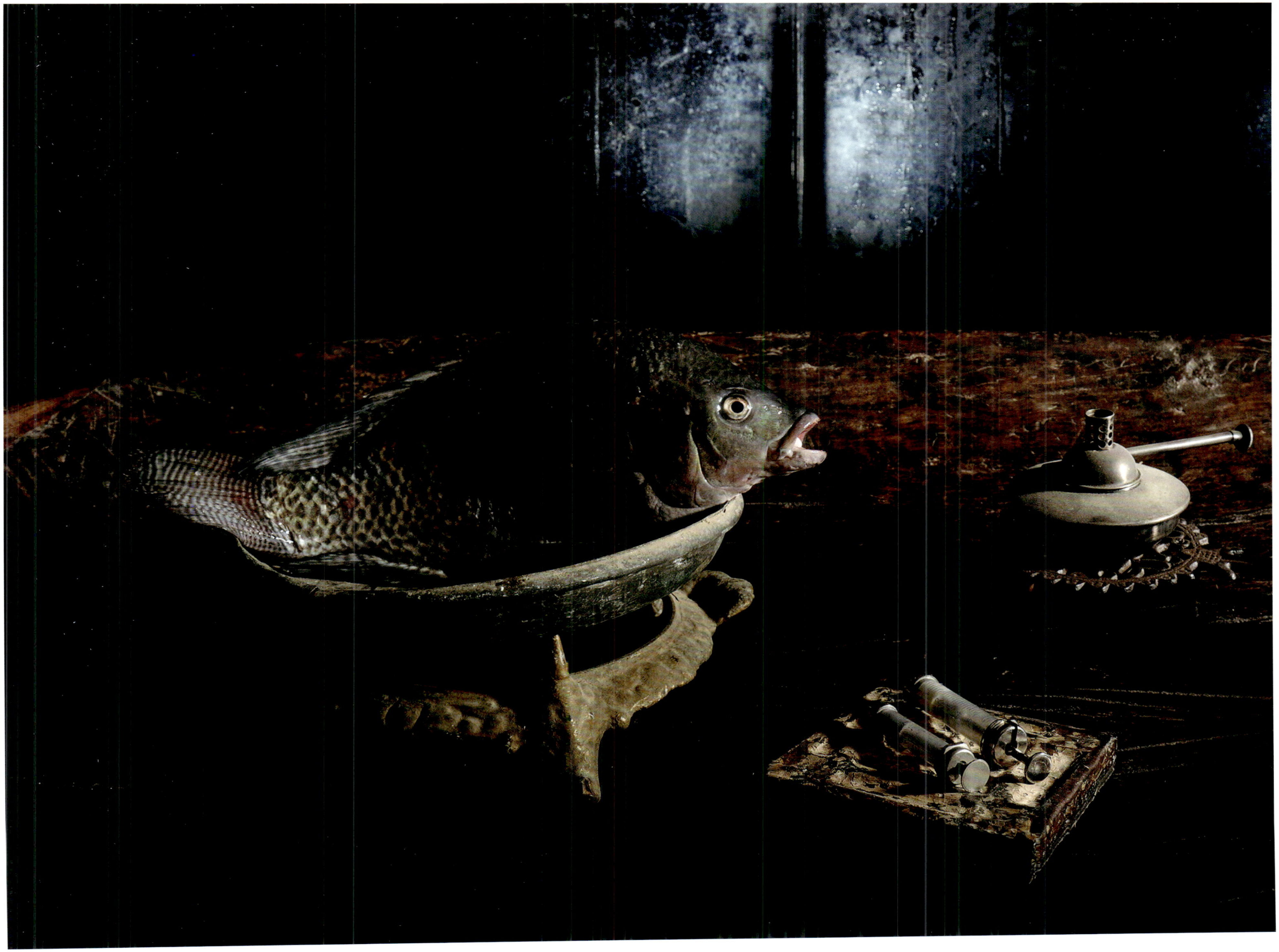

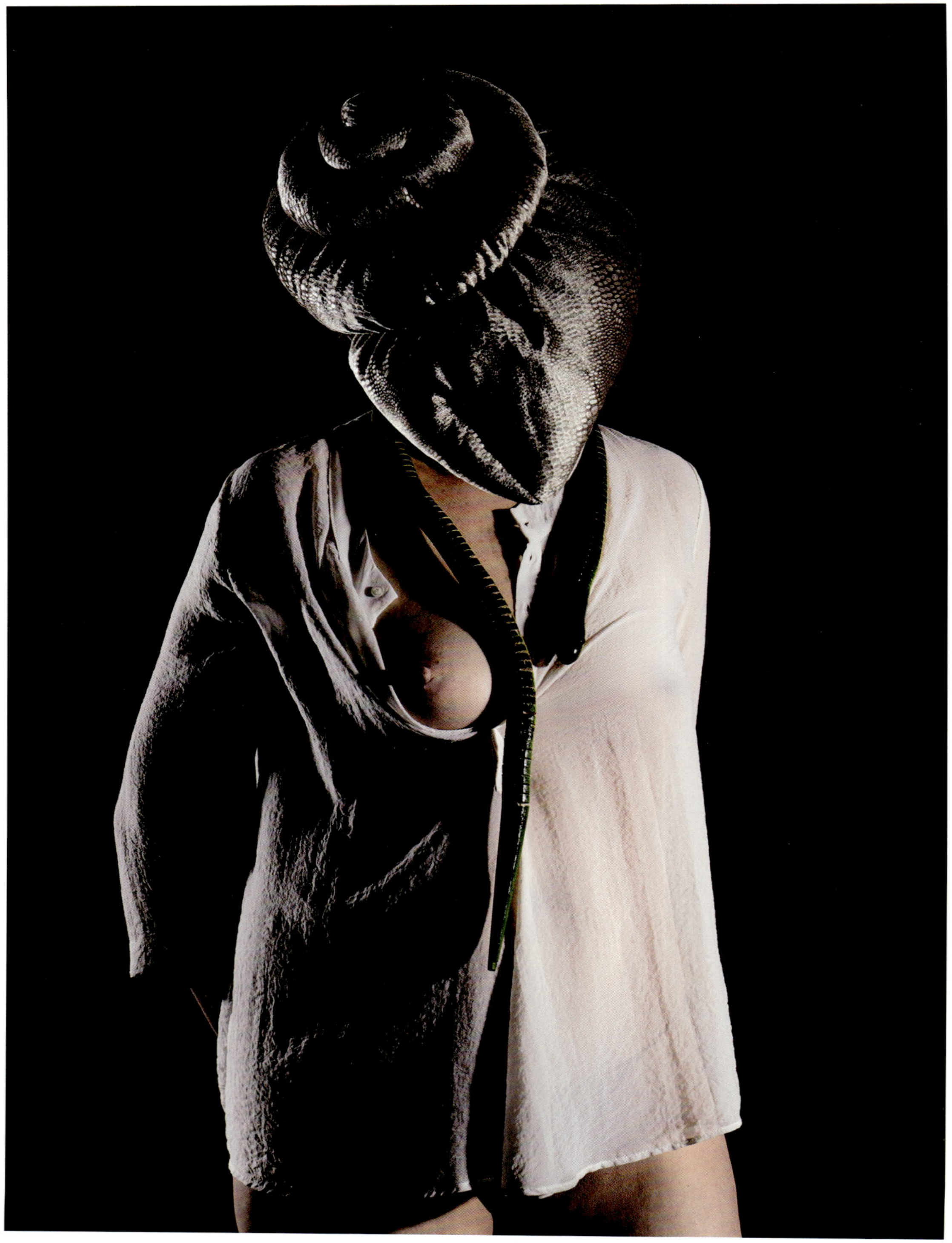

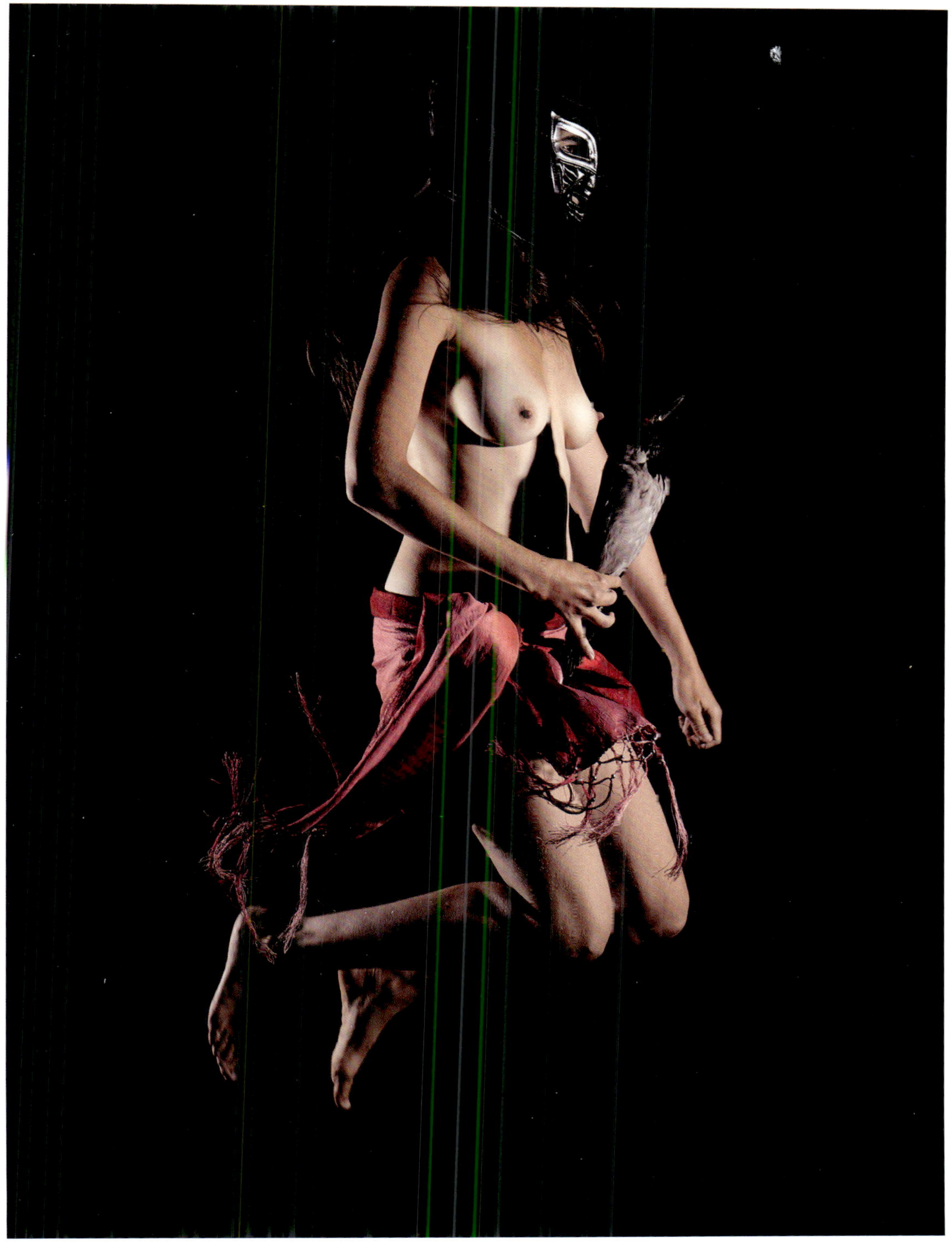

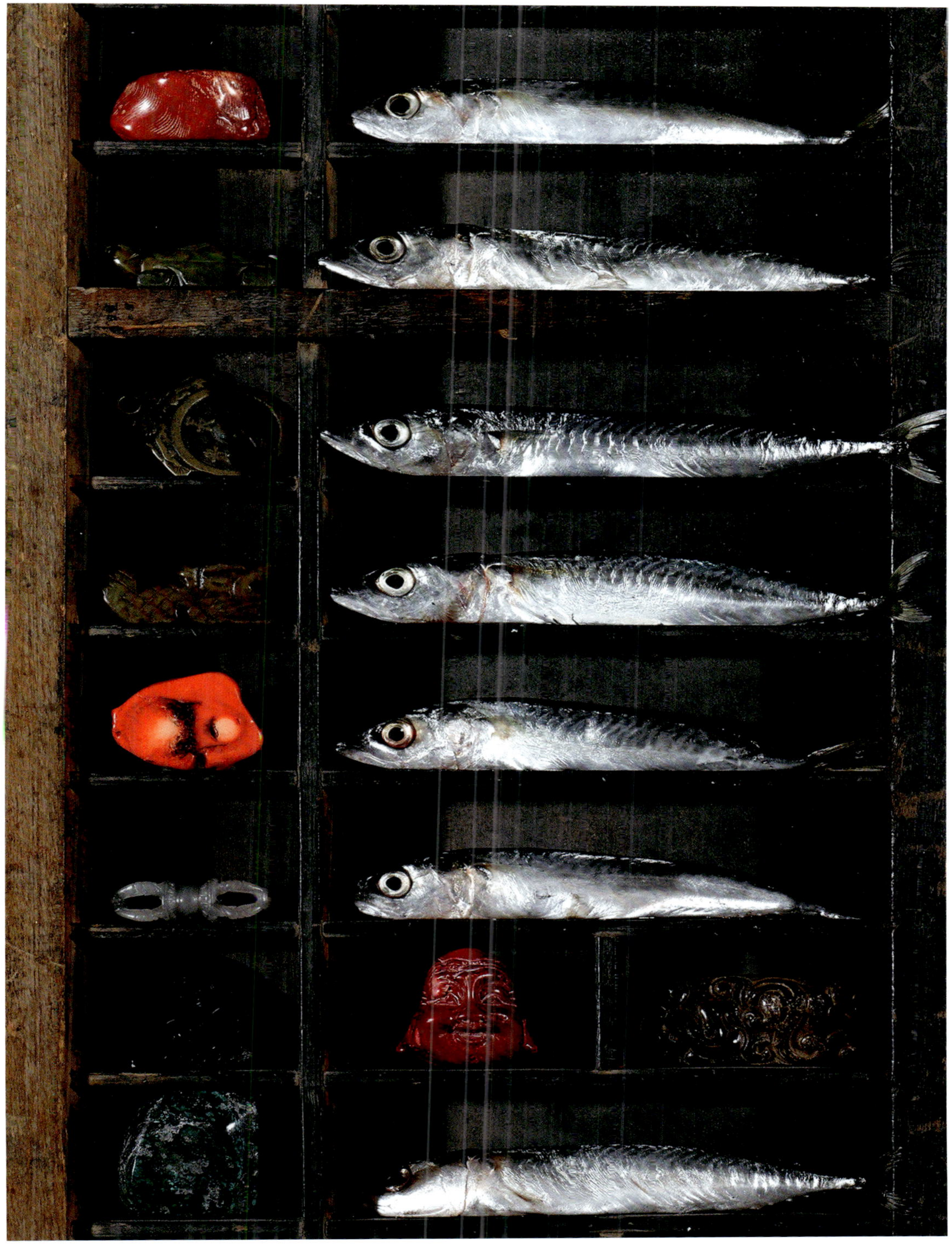

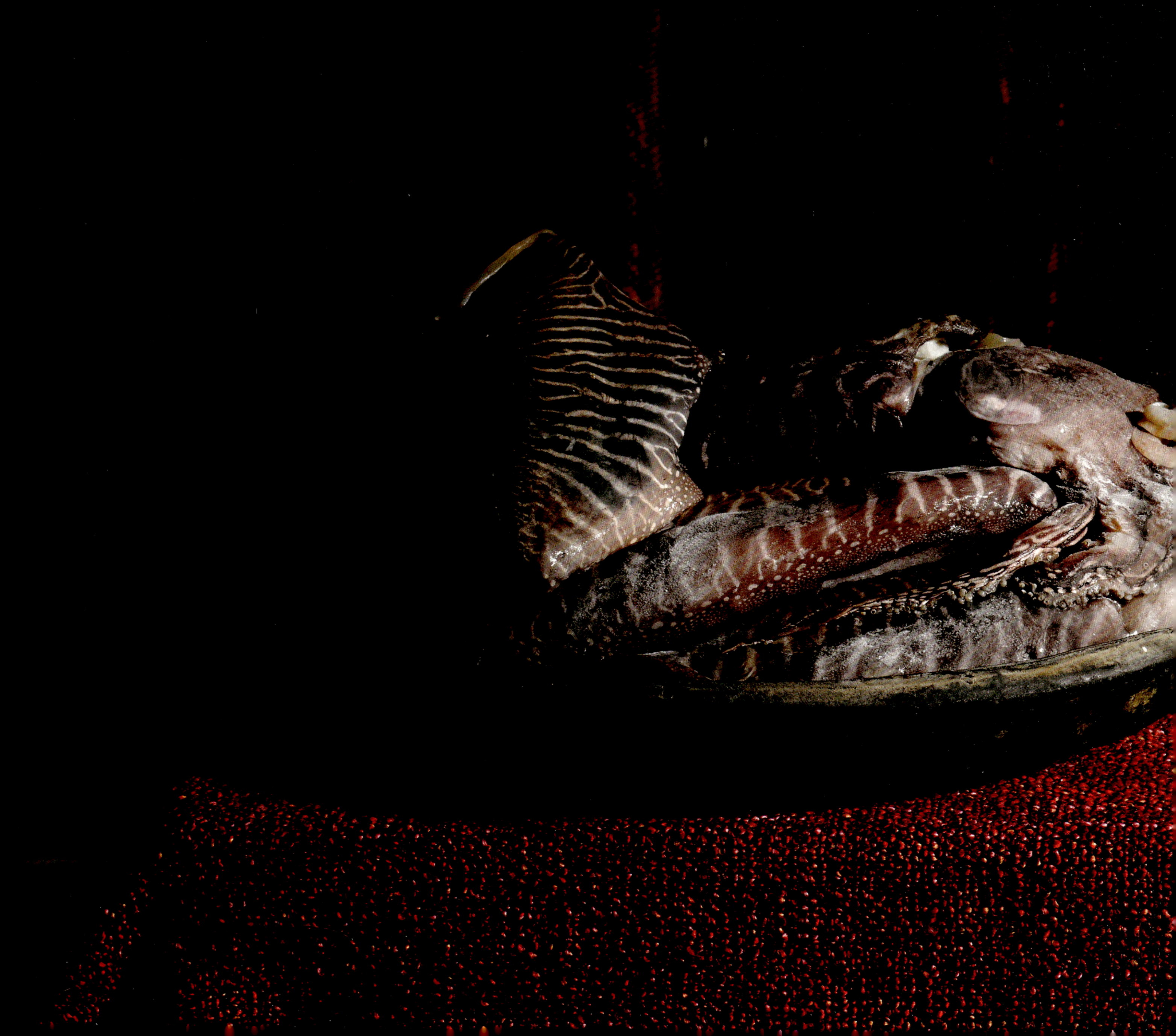

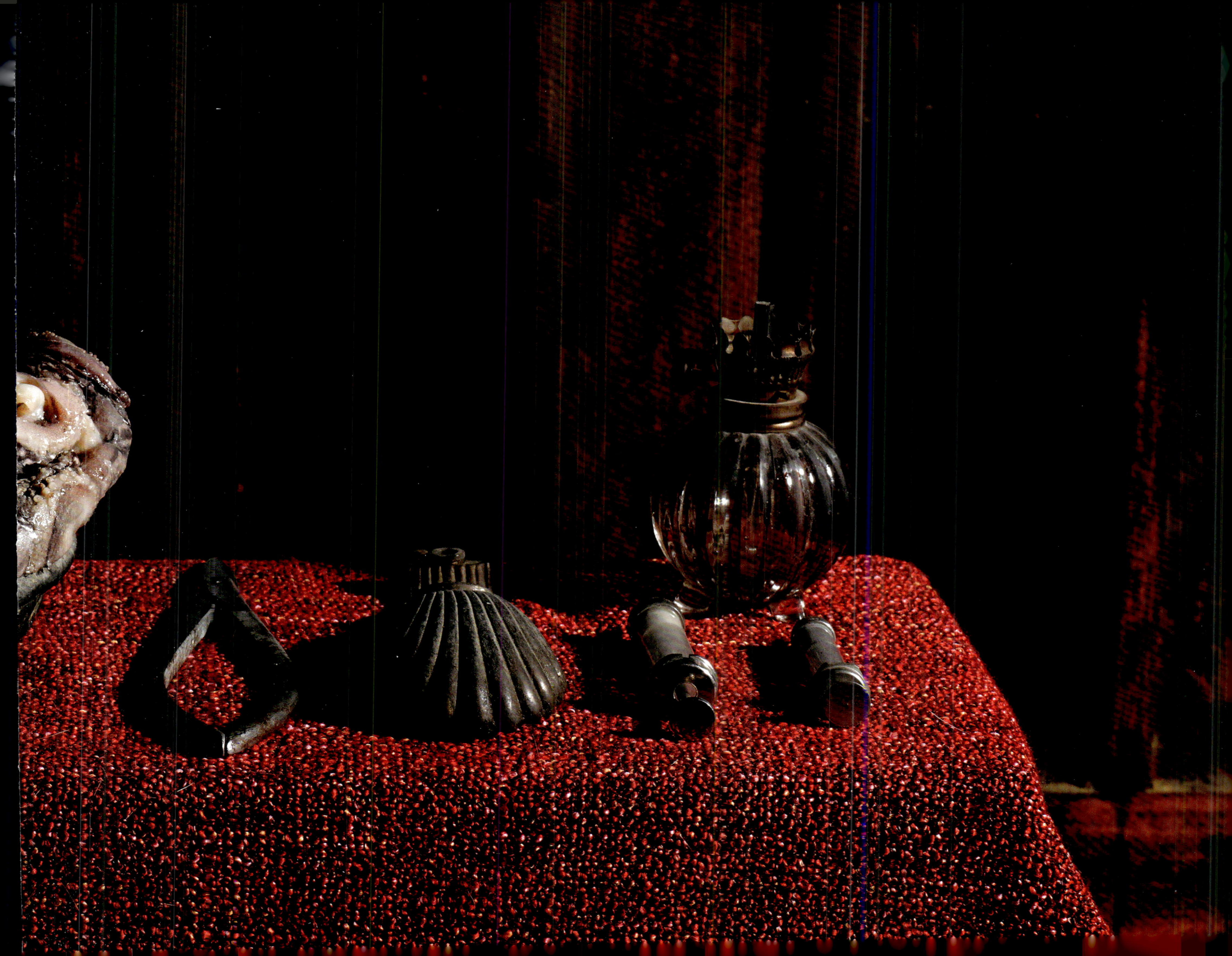

BIOGRAPHY

Eran Gilat – is a neuroscientist and avid fine art photographer. His scientific research focuses on the study of the mechanisms underlying epilepsy, as well as on the development of an innovative cure for this illness.

Eran's photographic work was introduced in a solo exhibition during the New York Photography Festival (2011) and received highly supportive critical responses. His works have been shown in leading galleries in Israel and abroad, including: FOTO8, London; LensCulture FotoFest Paris; Lodz Grand Prix Fotofestiwal; Gallery Huit, Arles; Philadelphia Photo Arts Center; Cape Town Month of Photography (MoP5, 2012); the San Diego Museum of the Living Artist; Athens Photo Festival 2012; Millennium Images (First Prize 2012); Edward Hopper House Art Center, Nyack, New York (Jury Recognition Award); the Cambridge University Gallery; the Photo Academy "Yanka Kurkchieva," Bulgaria; the On The Seam Museum, Jeruslame; the Griffin Museum of Photography, Boston; The Center for Fine Art Photography, Fort Collins, Colorado; NIKON Storytelling Award 2013, Tokyo; the Auckland Festival of Photography, New Zealand; "International Print Exhibition 157" Royal Photographic Society, Bath; "The Brain", New York Hall of Science (2014–15); Gera Museum, Serbia; PhotoPlace Gallery, Middlebury, Vermont); Head On Photo Festival 2015, Sydney, Australia; and the Rehovot Municipal Art Gallery, Israel (2016).

His photos have also been featured in numerous magazines and newspapers, including:
Sunday Times Spectrum, *GUP Magazine*, *Trend Tablet*, *Tribe Magazine*, *L'Oeil de la Photo*, *PH Magazine*, *LensCulture*, *Art.Es Magazine*, *Prism*, *BLUR Magazine*, *Archivo*, *SHOT! Magazine*, *fotomagazine.hu*, various South Korean photography magazines, *Musée Magazine*, *Slate*, *foc! photo magazine*, and *DIY Photography*.

Acknowledgments

Without the continuous support of Hana, Tom, Ron, and Ben, this project could not have been realized. I am sincerely indebted to Sam Barzilay, the gifted Creative Director of the New York UPI Gallery, for his interest in my activity, as well as for the initial and essential exposure of the project. I would also like to thank Klaus Kehrer and his team for their continous support in publishing this book.

Texts
Eran Gilat, Smadar Levy

Proofreading
Gérard A. Goodrow

Design and Image Processing
Kehrer Design Heidelberg

Production
Kehrer Design Heidelberg

Printed and bound in Germany
ISBN 978-3-86828-680-9

Kehrer Heidelberg Berlin
www.kehrerverlag.com